Fundamentals of
Biochemical Engineering

Fundamentals of Biochemical Engineering

Dr. A.V.N Swamy

M.Tech, Ph.D (IITB)
Professor, Chemical Engineering Department,
JNTU College of Engineering,
Anantapur - 515 002.

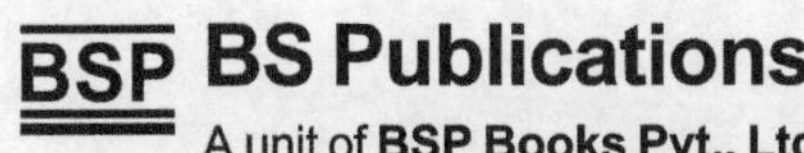

BS Publications

A unit of **BSP Books Pvt., Ltd.**

4-4-309/316, Giriraj Lane, Sultan Bazar,
Hyderabad - 500 095
Phone : 040 - 23445605, 23445688

Published by :

 **BS Publications**
A unit of **BSP Books Pvt., Ltd.**

4-4-309/316, Giriraj Lane, Sultan Bazar,
Hyderabad - 500 095
Phone : 040 - 23445605, 23445688
e-mail : info@bspbooks.net

ISBN : 978-93-52300-12-9 (HB)

*Dedicated to My
Beloved Parents*

JAWAHARLAL NEHRU TECHNOLOGICAL UNIVERSITY

DR. K. RAJAGOPAL
VICE-CHANCELLOR &
CHAIRMAN, EAMCET - 2006

Kukatpally,
Hyderabad - 500 072
Andhra Pradesh (India)
Grams : "TECHNOLOGY"
Phone : 040 - 23156109 (O) 23156116 (R)
Fax : 040 - 23165112
E-mail : rajagopalmail@rediffmail.com
www.jntu.ac.in.

Foreword

Process biotechnology is gaining importance due to it's increasing demand for producing various types of organic acids, enzymes, antibiotics, vaccines, and interferon etc. The Chemical Engineers need basic understanding of biology and biochemical mechanism, to develop the process for microbial products. The fundamentals of Biochemical Engineering is dealt in this book, for exposing the chemical engineering to basics of biochemical engineering.

I am glad to note that Prof. A.V.N. Swamy with his wide experience in process industries and experience in teaching biochemical engineering to B.Tech and M.Tech studies at J.N.T.U College of engineering has authored this book.

This book would be useful to undergraduate students pursuing B.Tech degree and postgraduate students pursuing M.Tech and M.Sc in biotechnology. The book encompasses the basic principles of biochemical engineering and would be useful in rendering clarity and in understanding the subject.

I am extremely happy to note that this publication would cater the needs of undergraduates and postgraducates of biotechnology.

I strongly recommend this book to the students of Chemical Engineering and Biotechnolgy and also to the faculty that deal with this subject at large. I wish the author all the success.

Prof. K. Rajagopal
Vice-Chancelllor,
Jawaharla Nehru Technology University (JNTU)
Kukatpally, Hyderabad.

Preface

This book is intended to cover most of the syllabus for Biotechnology and Biochemical Engineering students of undergraduate course. The need for a complete book which covers most of the important topics in Biochemical Engineering, prompted me to write this book. My keen interest in the subject of biochemical engineering and teaching experience at JNTU College of engineering, helped me to give a shape to this book. This book contains 15 chapters in total. Chapter 1 contains introduction to process biotechnology. Chapters 2 to 4 dealt with basic microbiology, like cell biology, metabolic pathways and microbial growth. In chapter 4, Monod kinetics, growth kinetics, continuous culture of microorganisms in stirred tank fermentors, have been sufficiently discussed for basic understanding of the subject. Chapter 5 discussed about various types of immobilization. A basic idea about immobilized enzyme reactors is introduced and control of enzyme activities is also dealt briefly. A brief mention is made about transport of molecules across cellular membranes, passive and facilitated diffusion is also discussed. Chapter 6 deals with classification of enzymes, and Kinetics of enzyme catalyzed reactions. Emphasis is made on Michaelis – Menten equations and estimation of Michaelis – Menten parameters. Enzyme inhibition is also discussed briefly. Bioreactor design is discussed in Chapter 7. Batch reactors, enzyme catalyzed reactors in CSTRs, CSTR with recycle with cell growth and ideal plug flow reactors, are discussed briefly in Chapter 7. Chapter 8 deals with media and bioreactor sterilization. Sterilization of media, sterilization of air and sterilization bioreactors are discussed. Thermal death of microorganisms and effect of temperature on specific death rate is discussed fairly in detail. Chapter 9 is focused on fermentation technology and industrial fermentation. Various commercial important industrial fermentations like production of penicillin, citric acid, enzymes and vaccines are discussed briefly. Also solid state and submerged fermentation is discussed to cover the basic concepts. Chapter 10 deals with aeration and agitation in bioprocess. Most of the industrial fermentations follow aerobic biological process. So it is essential to have basic understanding of aeration and transfer of air bubbles in the liquid media. Agitation is also important to properly mix the reactor contents, to obtain uniform concentration throughout reactor. Considering this aspects, stress is laid on oxygen transfer in liquid. The other important topics covered in this chapter include power requirements in aeration, factors effecting oxygen transfer coefficient, types of agitators and types of spargers. Efforts are made to cover various important separation process, normally deployed in bioprocess industries in chapter 11. To have basic idea of separation process, principles of separation are discussed in the beginning of this chapter. The separation process like centrifugation, filtration, and sedimentation have been discussed to have the fundamental knowledge of separation process. Disruption of cells, extraction and principles of chromatography are discussed along with ultrafilteration. The applications of such separation process and bioprocess industries, are discussed briefly. Chapter 11 gives an overview of various separation process practiced in various bioprocess industries. Scale up operation is discussed very briefly in chapter 12. Important principles of scale up are discussed initially to have an idea about scale up.

Various other points discussed in this chapter include scale up studies pertaining to power to unit volume and oxygen transferred coefficient are also discussed. The interesting part of this unit is introduction of basic concepts of scale down. Scale up with respect to aeration and agitation is also mentioned briefly. Chapter 13 deals with bioreactor instrumentation and control. Since control of a bioprocess is very important to maintain standards of bioproducts, instrumentation and control chapter is included as part of this book. Measurements during fermentation process, control fermentation are discussed briefly. Sensors of the physical environment, chemical sensors, analyses cell population are discussed as a part of process control. Chapter 14 discusses the principles of effluent treatment. Effluent treatment systems are becoming mandatory in bioprocess industries. This chapter deals only with biological treatment systems consisting of aerobic treatment systems, and anaerobic biological treatment. Also a brief mention is made about industrial waste water treatment. The medical applications of bioprocess engineering are discussed in chapter 15. Tissue culture process and gene therapy are discussed very briefly in this chapter to give a basic understanding. Stem cells have also been discussed, since stem cell research is gaining importances.

The contents of this book covers most of the syllabus covered for teaching Biochemical Engineering to undergraduate students of Indian Universities. The efforts are made to emphasis on application of basic biological principles to bioprocess engineering. I have put my sincere efforts to cover all the essential topics required for the basic understanding of biochemical engineering subject. This book also includes various figures at appropriate places along with photographs to aid students for comprehensive understanding of the subject. Review questions have been added at the end of each chapter for easy revision of each topic.

This book will be useful to undergraduate students of Chemical Engineering for their course in Biochemical Engineering and also to the students of Biotechnology who have biochemical engineering course.

I am thankful to my friends and colleagues for their good wishes and their supports. I am thankful to Prof. K. Rajagopal, Vice-chancellor, JNTU, Hyderabad, for writing the Foreword to this book. I would like to thank Mr.G.Sivarama Krishna for helping me during the course of preparation of this book. I thank Dr. V.K.Srinivas, General Manger R & D , M/s Bharat Biotech International Ltd, Hyderabad for providing relevant photographs to incorporate in this book.

Finally I dedicated this book to my parents for constantly encouraging me to do something always valuable.

I am grateful to the publishers, B.S. Publications, Hyderabad for bringng out this book after critical review to serve the undergraduate and post graduate students.

-Author

Contents

CHAPTER **1**

Basics of Process Biotechnology

1.0   Introduction

Bioprocess engineering in general helps in solving various problems and producing vital drugs for improving the mortality rate. Genetic engineering plays an important role in manipulating biological systems for their betterment. Various divisions like, genetic engineering, bioreaction engineering, bioassays, industrial microbiology, biochemistry, enzyme kinetics, upstream and down stream process do help in many ways for enhancing the life styles of human beings at large. Biochemical engineering is the confluence of microbiology, biochemistry and chemical engineering. Most of the microbiological based products are developed in the laboratory level and scaled up to the industrial level by following biochemical engineering principles. It is very essential to produce the products like, vitamins, vaccines, antibiotics, insulin etc, to meet the huge demand. A biochemical engineer plays an important role in developing the products from laboratory level to industry level. This book deals with the basic principles involved in biochemical engineering.

1.1   Bioprocess Engineering

Bioprocess engineering is required for developing various products by adopting, biotransformation, enzymatic reactions, various types of fermentations, selecting suitable media, selecting high strain, novel separation techniques. Obviously this being an interdisciplinary field the combined effort by microbiologists, biochemists, genetic engineers, physicists and chemical engineers is desired.

The scope is wide and a lot of efforts should be made to develop the bioprocess. Bioprocess engineering is being applied while producing high value, low volume biopharmaceuticals, insulin, interferon, enzymes and the like. Scientists and engineers should dream a lot for developing novel processes and technologies for producing products of importance. Particularly for combating various types of diseases like cancer, aids and infectious fevers like yellow fever. The penicillin story is the classic example for the team work of biologists and chemical engineers. After Alexander Fleming discovered penicillin in 1928, penicillin was used for treating a baby patient in London. The result was encouraging in treating the patient, however due to short supply of penicillin, the fever revoked and patient died. It was then felt to produce penicillin in large scale. Then microbiologists, biochemists and chemical engineers formed as a single group and produced penicillin in large quantities. Subsequently the demand for penicillin was increased during Second World War to treat the causalities. High yielding strains of *Penicillium chrysogenium* was isolated due to the remarkable work of dedicated scientists. Well known pharmaceutical companies like Pfizer and Merck could successfully produce large quantities of penicillin. Particularly Merck, USA achieved this task by assigning the problem to a group consisting of microbiologist, biochemist and chemical engineer. The consistent research and application of novel techniques resulted in enhancing the yield of penicillin in fermentation broth from 0.001 g/l to about 50 g/l.

Presently various reputed institutions like IITs and other institutions Like MIT, Purdue are engaged in high quality research.

Usage of proper microorganisms for carrying out a specific reaction of interest is very important. There are many culture collection centers all over the world. In India there are two culture collection centers namely Institute of Microbial Technology, Chandigarh and National Chemical Laboratory, Pune.

The significance of biotechnology is well recognized in India also and some of the educational institutes like HBTI, Kanpur, MUICT Mumbai, Anna university, Chennai are offering B.Tech and M.Tech courses. Institutes of higher repute like IIT Delhi, IIT Madras, IIT Bombay and IIT Kanpur are also offering M.Tech and PhD programs in process biotechnology.

Biotechnology based industries are also increasing in India since 1990s. Out of all the industries a few have reached reasonably good level in the international market namely M/s Biocon Industries, Bangalore, Bharat Biotech Ltd., Hyderabad and Shanta Biotech, Hyderabad. A few biotechnology based industries in India are given in Table 1.1.

Table 1.1 Partial list of Indian industries producing biotechnology based products.

S.No.	Company	Product
1.	Biocon Industries, Bangalore	Enzymes
2.	Bharat Biotech, Hyderabad	Hepatic B vaccine
3.	Shanta Biotech, Hyderabad	Heaptic B vaccine
4.	Krebs biochemicals, Srikakulam, Nellore	Statins, Effidrin
5.	Malladi Biotech, Mahad	Effidrin
6.	Aurobindo pharmaceuticals	Semisynthetic Penicillins
7.	Lupin chemicals, Mumbai	Cephalosporines.
8.	Hindustan antibiotics Ltd, Pimpri, Pune	Penicillin G
9.	IDPL Rishikesh	Penicillin G

Various biotechnology based products have been added continuously as a result of active academic and industrial research. The CSIR laboratories like IICT, Hyderabad, National Chemical laboratory, Pune and CFTRI, Mysore are also involved in active research in biotechnology. State Governments in India are establishing biotechnology parks for promoting industrial research.

The products that can be produced by adopting biotechnology are given in the Table 1.2

Table 1.2 Biotechnology based products.

S. No.	Product
1.	Vitamins B-2 (Riboflavin) ; B-12 (Cyanocobalamine) ; C (ascorbic acid)
2.	Amino acids L-lysine , L-glutamic acid
3.	Nucleic acids
4.	Organic acids Citric acid, acetic acid, lactic acids, gluconic acids
5.	Antibiotics Penicillin G, semisynthetic pencillins, Cyphalosporines, tetracyclines
6.	Vaccines Hepatic B, C, and A
7.	Human insulin
8.	Human interferon
9.	Industrial enzymes Protease, Amylase, Kinase, Q-enzyme

1.2   Role of Biochemical Engineer in Bioprocess Industries

Biochemical engineer's role in bioprocess industries is very important. Various microbial products have been developed at research laboratory level and these products should be commercialized without losing the yield and concession. Also new type bioreactors may be tried for improving the product quality. For example during penicillin production mycelia are formed in the stirred tank bioreactor eventually the viscosity of the fermentation broth is increased. These results in enhancement of power required for unit reactor volume. This problem may be solved by using fluidized bed reactor where the immobilized whole cells are fluidized. Some of the unit operations which are to be studied by biochemical engineers are, mixing of heterogeneous phases, microorganisms, medium and air. Also the areas of interest for further studies are, mass transfer of oxygen from air to microorganisms through liquid, heat transfer from reacting liquid etc. Biochemical engineer has to study the aspects of various unit operations involved in bioprocess like oxidations, substrate conversions, biotransformation, hydrolysis, biopolymerisation, biosyntheses and growth of microorganisms.

The process design of various bioprocesses needs the study of bioreactor engineering, thermodynamics, stoichiometry etc. Most of the bioprocesses are carried out in batch process. And study of continuous process is being carried out as a part of active research. Enzymes can be produced either by solid state fermentation or by submerged fermentation.

Depending on the yield and conversion, suitable process may be selected after thoroughly conducting experiments on both types of processes. Computer controlled bioreactors are useful in understanding the fate of various process parameters during the fermentation like dissolved oxygen, carbon dioxide, redox potential and glucose levels. Certain variables like power, aeration and substrate utilization can be studied with the help of sophisticated bioreactors.

The various products that can be produced by utilizing microorganisms of various species are represented in the Fig. 1.1.

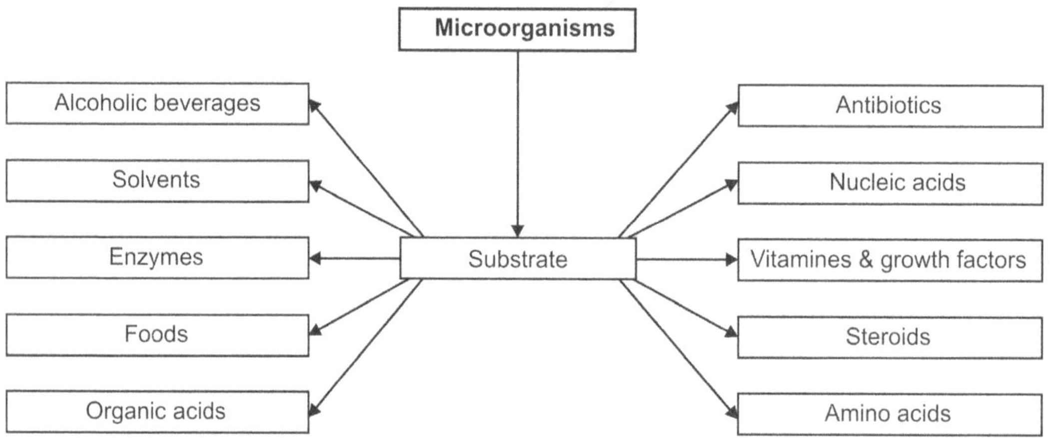

Fig. 1.1 Various products produced by using microorganisms.

The development of bioprocesses involves continuous experimental work which gives reproducible results. Since the complete process biotechnology is a trade secret, it is inevitable to develop the process technology indigenously. The latest bioassay

Techniques available will facilitate to understand the nature of the product produced through bioprocess technology.

1.3   General Bioprocess Overview

The bioprocesses have some stages which are common during production of microbial products. These stages are given in the Fig. 1.2. The details of the processes will be discussed in subsequent chapters.

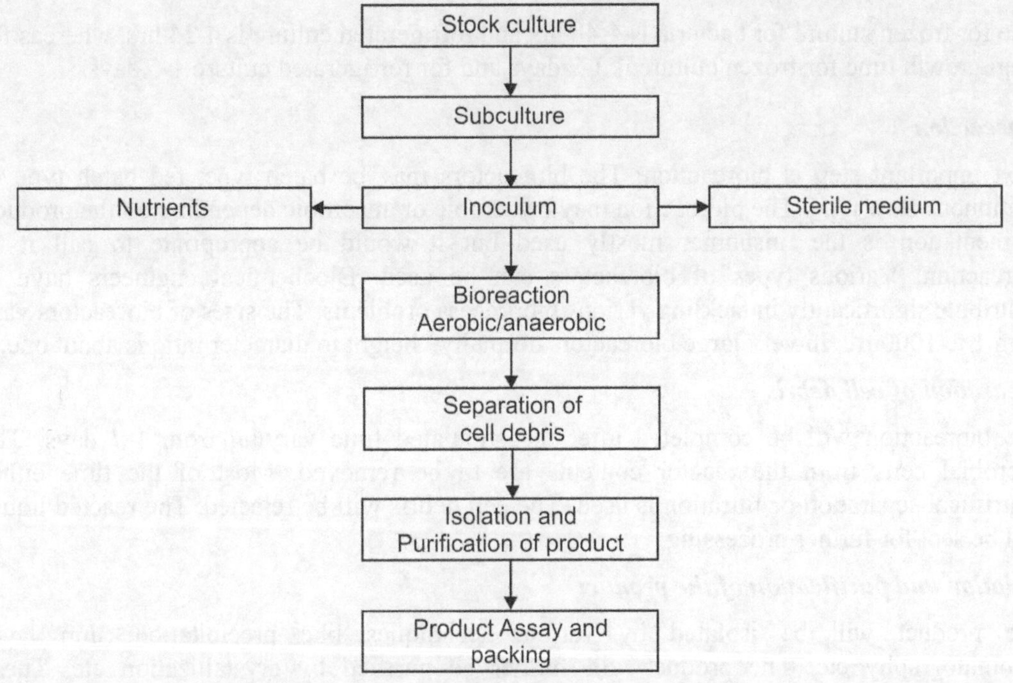

Fig. 1.2 Various stages of a general bioprocess.

Stock culture

The strain which is responsible for a particular bioprocess has to be identified and procured from the culture collection centers. The culture obtained from such collection centre should be preserved carefully without any contamination. The established microbial culture which gives optimum yield and conversion is the stock culture used for the production of specific product. The microbial cultures are stored at low temperatures for short term storing, temperatures of below 5 ^{o}C are used, for long period of preservation – 18 ^{o}C to – 80 ^{o}C are used, for storing for very long periods lyophilisation is used.

Subculture

Subculture has to be developed from the stock microbial culture, by using standard methods. This has to be prepared for every batch, so that the stock culture is always available.

Inoculum

Inoculum will be produced from the subculture. The required quantity of culture will be prepared depending on the reactor liquid volume. In some distilleries this is called as propagation of culture. The microbial culture is initially developed in shake flasks and further taken into larger vessels as per the requirement of commercial bioreactors. This Inoculum will be normally 15 to 20 % of bioreactor volume.

The microbial culture has to be maintained at optimum growth prior to inoculating to bioreactors. The growth time required for lyophilized cultures is about 3-14 days. The growth time for frozen culture for bacteria is 4-48 hrs and refrigerated culture is 4-24 hrs, whereas for fungi growth time for frozen culture is 1-7 days and for refrigerated culture 1-5 days.

Bioreaction

Next important step is bioreaction. The bioreactors may be batch type, fed batch type or continuous flow type. The bioreaction may be aerobic or anaerobic depending on the product. Fermentation is the misnomer mostly used but it would be appropriate to call it as bioreaction. Various types of bioreactors can be used. Biochemical engineers have to contribute significantly in tackling various bioprocess problems. The sizes of bioreactors vary from 1 to $1000m^3$. In very large bioreactors normally   height to diameter ratio is about one.

Separation of cell debris

The bioreaction will be completed after the stipulated time varying from 1-7 days. The microbial cells from the reactor contents are to be removed. Most of the time either centrifugal separation or filtration is used. The cell debris will be rejected. The reacted liquid will be sent for further processing.

Isolation and purification of the product

The product will be isolated by various techniques like precipitation, thin layer chromatography etc. The product will further be purified by crystallization etc. These procedures will be discussed in detail later.

Product Assay

The product will be tested by the international standard methods for confirming their purity. And finally the product will be packed.

Effluent Treatment

Finally the effluent generated, after recovering the product, will be treated either by biological treatment or by chemical treatment.

References

1. Aiba, S.Humphrey, A.E. and Millis, N.F.(1973), Biochemical Engineering, Academic Press, New York.
2. A.H.Patel, (1984) Industrial Microbiology Mac Millan India Ltd, Delhi.
3. D.G.Rao (2005) Introduction to Biochemical Engineering, Tata McGraw-Hill Publishing Company Ltd, New Delhi.
4. Michael J. pelzar, Jr, E.C.S.Chan, Noel R. Krieg (1996) Microbiology, Tata McGraw Hill Ltd., New Delhi.
5. Michael L.Schuler, Fikret Kargi (2002) Bioprocess Engineering basic concepts, Prentice Hall of India Pvt.Ltd., New Delhi.
6. Stanbury, P.F. and Whitaker, A. (1997), Principles of Fermentation Technology, Aditya Books (P) Ltd., New Delhi.

Review Questions

1. What is bioprocess engineering ?

2. What are the various products produced by using process biotechnology ?

3. Explain various stages of any bioprocess.

4. What is the role of a biochemical engineer in a bioprocess industry ?

5. What are the optimum temperatures required to store microorganisms ?

CHAPTER 2

Fundamentals of Microbiology

2.1 Biophysics and Cell Doctrine

Microbiology is the study of tiny living organisms which cannot be seen by an eye. Some of the organisms that fall into this category include bacteria, fungi, algae, protozoa and the like. Microbiology deals with their structure, classification, reproduction and physiology, the study of their distribution, their relationship with each other and their effects on plants and animals.

Microorganisms are utilized for production of useful products like cheese, alcohol, yogurt, wine, penicillin and interferon. The microorganisms are single cellular in nature and all the biological reactions are carried out by a single cell. In higher organisms the cells are formed into tissues and organs and carry–out the specific functions. There are various types of cellular organisms and these cellular organisms are categorized basically into two types, i.e., prokaryotes and eukayotes. The structure of the cells became vivid after the invention of electron microscope. Most of the cells fall under prokaryotic or eukaryotic category.

Procaryotes

The nucleus of the prokaryotic cells is not enclosed with well defined membranes. The prokaryotic cells are small in size and they are found alone instead of association with other cells. They are available in various shapes like, spherical, rods or spiral shape and sizes vary from 0.5µm. The volume of prokaryotic cells is about 10^{-12} ml per cell and about 80% is water. The mass of single prokaryotic cell is about 10^{-12} g. The doubling time of procaryotes is about 20 min and they take variety of nutrients for their survival. Because of quality of adoptability to nutrients from environment, and rapid growth, these organisms are used in research and bioprocess by different biotechnology based industries.

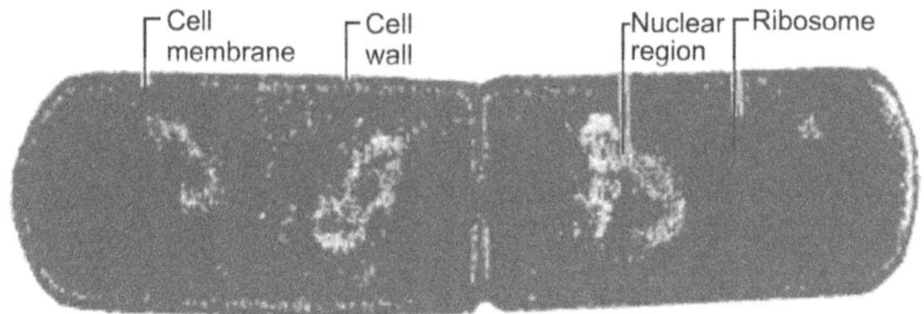

Fig. 2.1 Electron micrograph of a prokaryote.

The cell wall thickness is about 200 A^o and the cell membrane beneath this has thickness 70 A^o. The cell does not possess defined region like nuclear zone. The cell membrane is selective in transferring components between cell and the environment. The cell has ribosomes that look like dark spots in the cell. Some procaryotes like blue green algae carryout photosynthesis in the presence of sunlight.

Eukaryotes

These are large in number. These eukaryotic cells have a nucleus surrounding by the well defined membrane. The size of the eukaryotes is many more times larger than the procaryotes, may be more than 1000 times. The cell structure of eukaryotes is complex and the cell structure includes well distinct compartments. The cell structure is shown in Fig. 2.2.

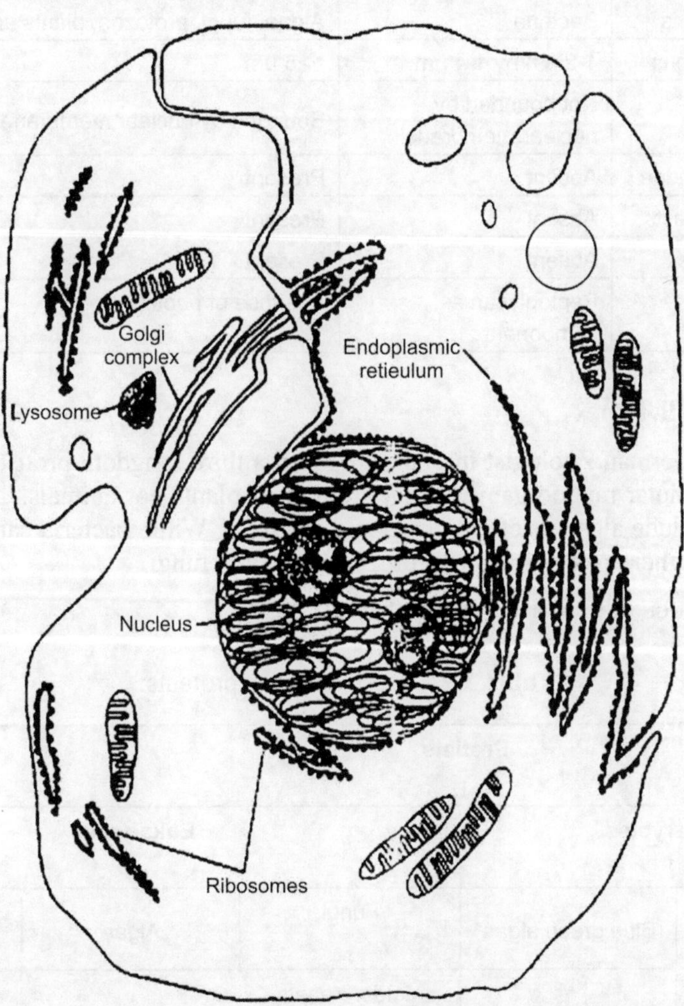

Fig. 2.2 The cell structure of eukaryotic cell.

This cell is entirely covered by cell membrane. Endoplasmic reticulum a convoluted membrane spreads from cell membrane into the cell. The endoplasmic reticulum contains ribosomes, the biochemical reaction sites. The catalytic activity at the ribosomes is controlled by the nucleus. Mitochondrias are the organelles major energy suppliers of the cell. Mitochondria use oxygen for energy generation in the cell. The other components of the cell include golgi complex, lysosomes and vacuoles and their main function is to isolate chemical reactions or chemical compounds from cytoplasm. The basic features of prokaryotic cells and eukaryotic cells are given in the Table 2.1.

Table 2.1 Basic features of prokaryotic cells and eukaryotic cells.

S.No.	Description	Prokaryotic cells	Eukaryotic cells
1.	Organisms	Bacteria	Algae, fungi, protozoa, plants animal cells
2.	Dimensions	1-2 µm by 1-4 µm	>>5 µm
3.	Nucleus	Not bounded by nuclear membrane	Bounded by nuclear membrane
4.	Mitochondria	Absent	Present
5.	Chloroplasts	Absent	Present
6.	Vacuoles	Absent	Present
7.	Cell wall	Peptoglycan as component	Absence of peptoglycan

2.2   Major Cell Types

E.H. Haeckel a German Zoologist in 1866 proposed a third kingdom protista. This kingdom consists of unicellular microorganisms that are neither plants nor animals. The examples for these protests include algae, bacteria, fungi and protozoa. While bacteria can be classified as lower protests, higher protests include protozoa, algae and fungi.

The classification of protits is given in Table 2.2.

Table 2.2 Classification of protests.

Protists				
↓				
Procaryotes			Eukaryotes	
↓			↓	
Bacteria	Blue green algae	Fungi ↓	Algae	Protozoa
		Molds, Yeasts		

The classification is basically made on the characteristics of energy and nutrient requirements, growth rates, reproduction, and motility.

Bacteria

The surface area / volume ratio of bacteria is very high when compared to larger organisms of similar shape. The shape of the bacterial cell may be spherical cocci, straight rods bacilli or helical spirella as shown in Fig. 2.3. The outer surface of bacteria is covered with the sticky substance called slim layer.

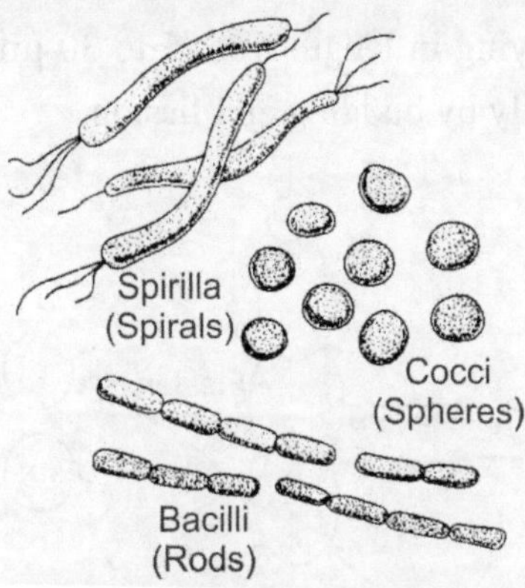

Fig. 2.3 Forms of bacteria.

The bacteria are capable of motion and they reproduce by binary fission. Characteristics of some bacteria are given Table 2.3.

Table 2.3 Characteristics of some bacteria.

S. No	Microorganism	Form	Gram test	Aerobic
1.	Acetobacter	Rods	- ve	Yes
2.	Bacillus	Rods	+ ve	Yes
3.	Chlostridium	Rods	+ ve	No
4.	Corenybacterium	Irregular form	+ ve	No
5.	Lactobacillus	Rods or cocci	+ ve	No
6.	Pseudomonas	Rods	-ve	Yes

Bacteria that retain blue crystal violet color after gram staining is called gram positive and that lose color is known as gram negative. In aerobic process bacteria utilize oxygen. Most of the commercial processes are aerobic, like organic acids, antibiotics, vinegar etc. Bacteria have the capability to form endospores in unfavourable environment. The bacteria can survive boiling in water for several hours. Spores are dormant in nature and they can resist heat, radiation and chemical environment. So the elimination of bacteria can be achieved by subjecting them to 12 psig and temperatures about 120 ^{o}C.

Fungi : Yeasts

Yeasts like bacteria are wide spread in the environment. Yeasts are available as single and small cells and their size is varying in length from 5 to 30 μm and in width 1 to 5 μm. The reproduction in yeasts is normally by budding and fission.

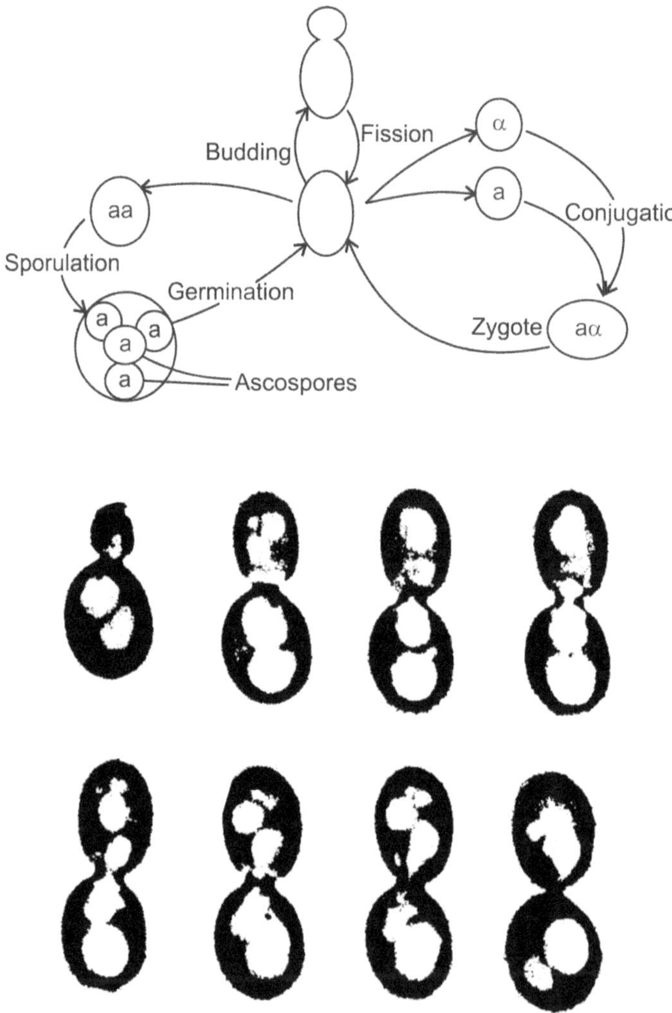

Fig. 2.4 Yeast reproduction by budding.

The other way of reproduction in yeasts is by binary fission. Sexual reproduction takes place by configuration of two haploid cells. The zygote thus formed may undergo several divisions to form ascospores. Each ascospore then becomes a new cell. The new cell may further undergo by asexual production by budding or by fission to produce cells. Yeasts have wider applications in industries like, beer production, wine, industrial alcohol and glycerol production.

Fungi: molds

The vegetative structure is found in higher fungi like molds.

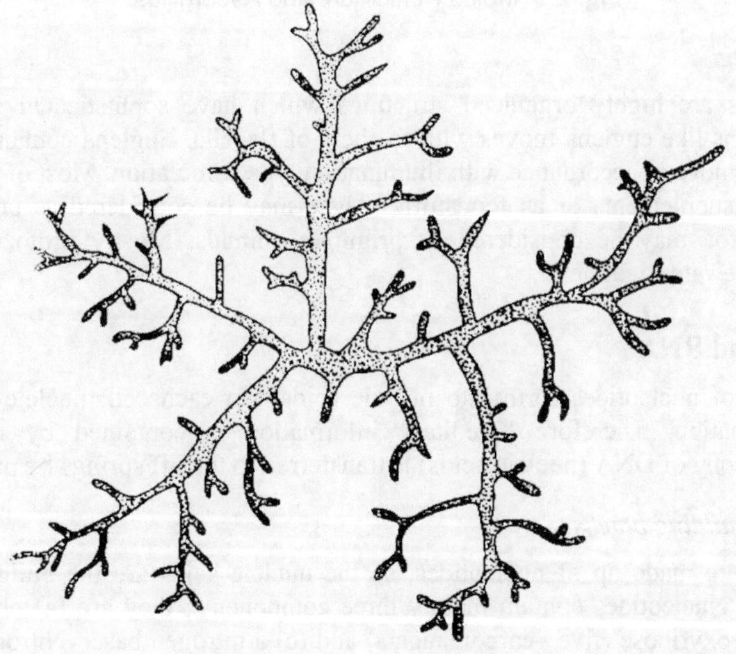

Fig. 2.5 Mycelium of molds.

The branched mycelia contain tube like structure as shown in the Fig. 2.5. The tubes contain mobile mass of cytoplasm. This cytoplasm contains many nuclei. Hyphae are the long filaments of cells contained in the mycelium. Molds are generally non-motile in nature. The molds of commercial significance are *Penicillium* and *Aspergillus*.

Citric acid is produced by *Aspergillus niger* by limiting phosphate and some metals like Fe and Mn.

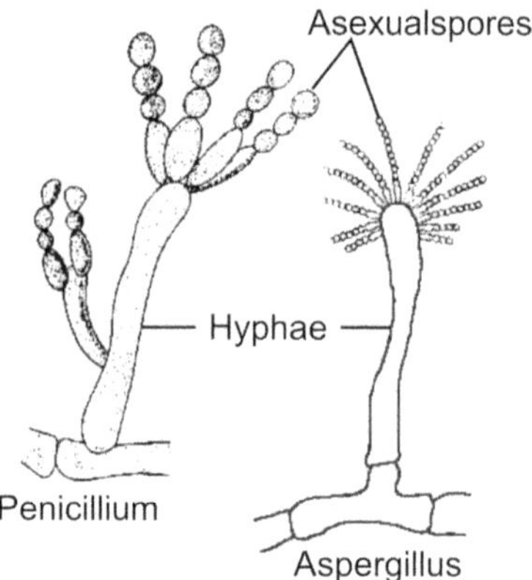

Fig. 2.6 Molds Penicillium and Aspergillus.

Algae and Fungi

The eukaryotes are highly organized structures which have sophisticated structure. The higher organisms like euglena moves with the help of flagella. Euglena contains an eye spot which helps to move in accordance with illumination type stimulation. Most of algae are used either as food supplements or as foodstuffs. Algae may be considered as primitive plants whereas protozoa may be considered as primitive animals. Mostly protozoa have been utilized in wastewater treatment.

2.3   DNA and RNA

The polymers of nucleotides form into nucleic acids. In each cell, nucleic acids contain genetic information. Therefore hereditary information is contained by nucleic acids. The complete copy of DNA (nucleic acids) is transferred to the off springs by parent cells.

Nucleotides: Building blocks

Nucleic acids are made up of nucleotides, so the nucleic acids are the building blocks of nucleic acids. Nucleotides contain mainly three components; they are (a) phosphoric acid (b) ribose or deoxyribose (five –carbon sugars) and (c) a nitrogen base. Nitrogenous base is derived from pyrimidine or purine.

Purine   Pyrimidine

The three components mentioned above are joined in a similar fashion in two nucleotides i.e., deoxyribonucleotides and ribonucleotides. They are distinguished by the five carbon sugar they possess.

RNA and DNA are biopolymers of ribose containing nucleotide and deoxyribose containing nucleotides respectively.

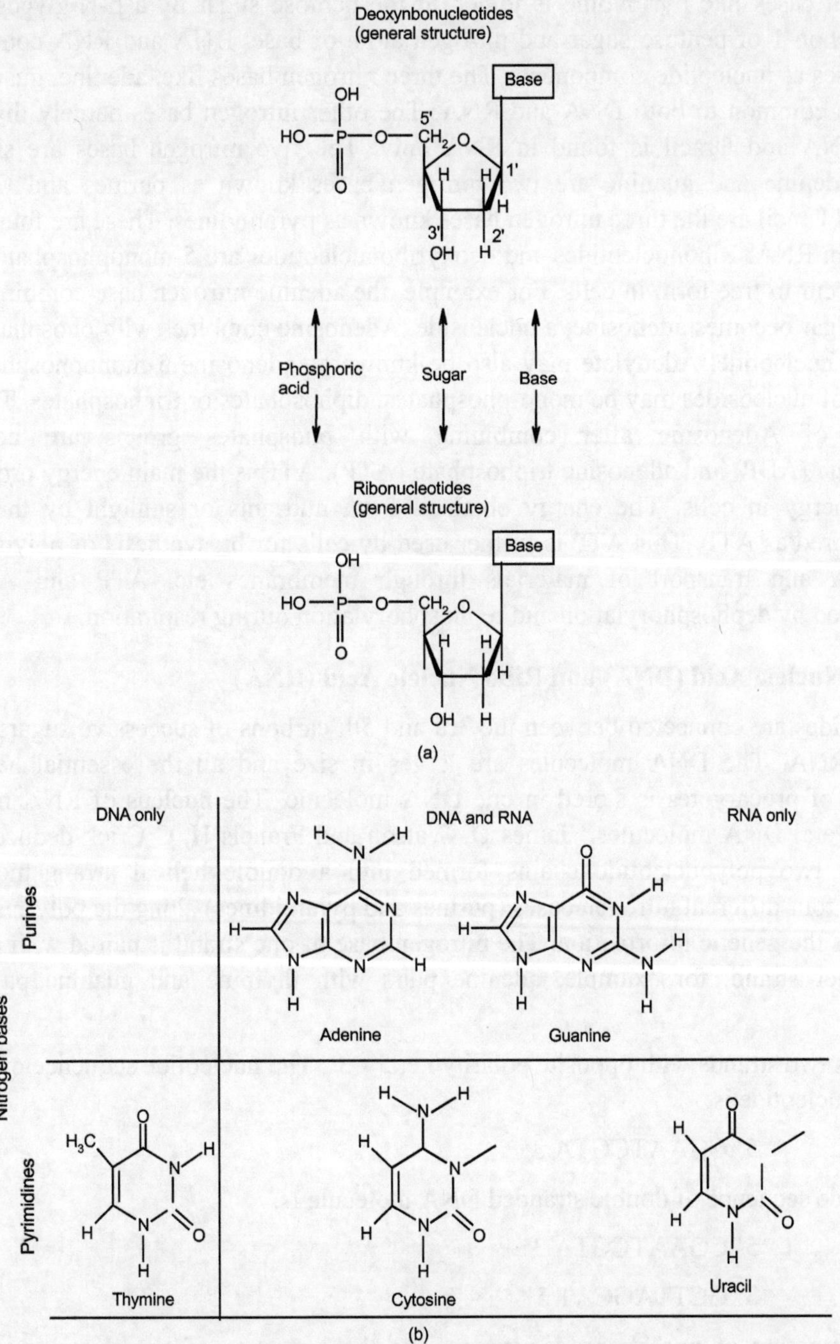

Fig. 2.7 The general character (a) Ribonucleotides (b) Deoxy Ribonucleotides.

Nucleotide consists of nucleoside and phosphate. The phosphate is joined to pentose of the nucleoside at the carbon 5 position. Nucleosides consist of nitrogen base and pentose sugar. The nitrogen bases like pyrimidine is joined to the pentose sugar by a β-N-glycosyl bond between carbon 1 of pentose sugar and nitrogen atom of base. DNA and RNA contain five nitrogen bases as nucleotide components. The three nitrogen bases like, adenine, guanine and cytosine are common to both DNA and RNA. The other nitrogen bases namely thymine is found in DNA and Uracil is found in RNA only. The five nitrogen bases are shown in Fig. 2.7. Adenine and guanine are two nitrogen bases known as purines and Cytosine, thymine and Uracil are the three nitrogen bases known as pyramidines. These are found either in DNA or in RNA. Ribonucleotides and deoxyribonucleotides are 5-monophosphates which generally occur in free form in cells. For example, the adenine nitrogen base combining with a pentose sugar becomes adenosine, a nucleoside. Adenosine combines with phosphate forms adenylate a nucleotide. Adenylate may also be known as adenosine-5 monophosphate. The derivatives of nucleosides may be mono-phosphates, diphosphates or triphosphates. The other derivatives of Adenosine after combining with phosphates groups are adenosine 5-diphosphate (ADP) and adenosine triphosphate (ATP). ATP is the main energy provider of chemical energy in cells. The energy obtained from nutrients or sunlight by the cell is normally stored as ATP. This ATP is further used by cells for biosynthesis of polymers like DNA, RNA and transport of materials through membranes etc. ATP and ADP are interconverted by dephosphorylation and rephosphorylation during respiration.

Deoxyribo Nucleic Acid (DNA) and Ribo Nucleic Acid (RNA)

The nucleotides are connected between the 3rd and 5th carbons of successive sugars both in DNA and RNA. The DNA molecules are larges in size and all the essential hereditary information of procaryotes is stored in one DNA molecule. The nucleus of RNA molecule contains several DNA molecules. James D. Watson and Francis H, C Crick deduced DNA structure as two polynucleotide chains formed into a double helical arrangement. The sequence of four different nitrogen bases (purines and pyramidines) along the poly-nucleotide chain carries the genetic information. The nitrogen base of one strand is paired with nitrogen base of other strand, for example, adenine pairs with thymine and guanine pairs with cytosine.

DNA has two strands with opposite polarity i.e. 5'- 3'. The nucleotide sequence of single Deoxyribonucleotide is,

5' CGAATCGTA 3'

Nucleotide sequence of double stranded DNA molecule is,

5' CGAATCGTA 3'

3' GCTTAGCAT 5'

The information of one strand sequence gives the sequence of complementary strand. So each strand is template of other.

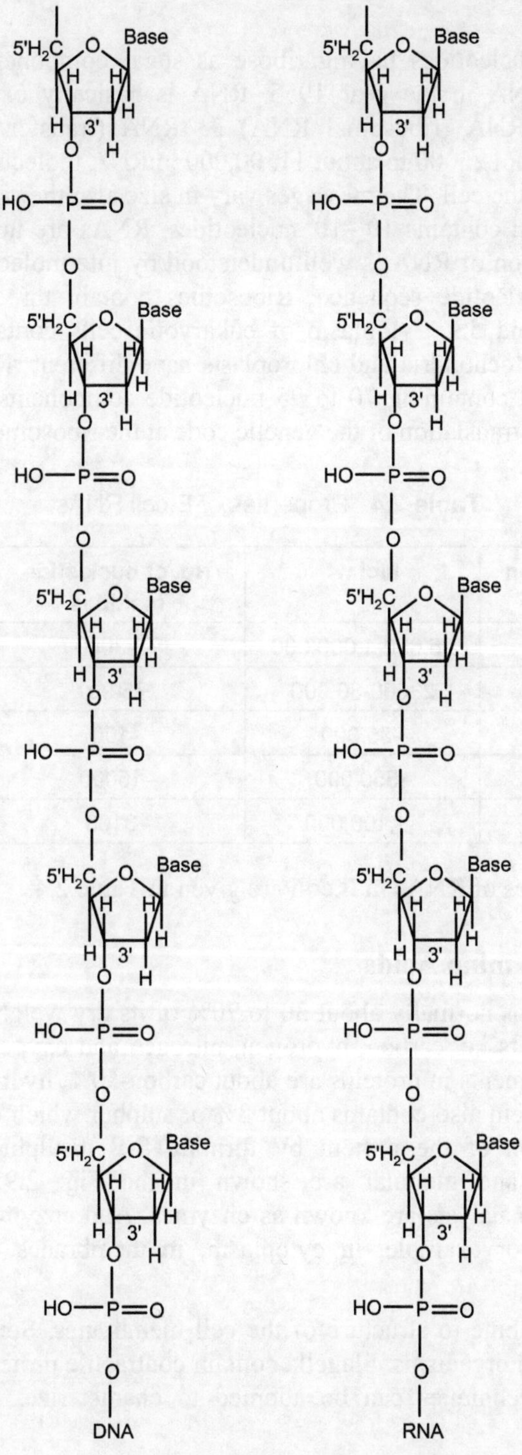

Fig. 2.8 Structure of DNA and RNA.

RNA

RNA consists of ribonucleotides having ribose as sugar component. Frankel Conrat and Williams discovered RNA in the year 1955. RNA is basically of three types.-1. mRNA (messenger RNA) 2. rRNA (ribosomal RNA) 3. tRNA (transfer RNA). The molecular weights of RNA vary from 25000 to about 11, 00,000. mRNA molecule carries message from DNA to another part of the cell. The messages vary in size also the sizes of mRNA vary. The typical length of a chain contains 10^3–10^4 nucleotides. RNAs are normally single stranded. The structure and function of RNA is well understood by intermolecular and intramolecular base pars of RNA nucleotide sequence. Ribosomes contain three different rRNAs and denoted as 23S, 16S and 5S. Cytoplasm of eukaryotic cells contains ribosomes and the ribosomes present in mitochondria and chloroplasts have different rRNAs. tRNAs are found in cytoplasm of the cell containing 70 to 95 nucleotide components. The main function of tRNAs is to assist in the translation of the genetic code at the ribosome.

Table 2.4 Properties of E.coli RNAs.

Type	Sedimentation coefficient	Mol wt	No. of nucleotide residues	Percent of total cell RNA
mRNA	6S-25S	25,000-1,000,000	75-3000	–2
tRNA	–4S	23000-30,000	75-90	16
tRNA	5S	–35,000	–100	82
	16S	–550,000	–15000	
	23S	–1,100,000	–3100	

Various characteristics of RNAs in E.coli are given in Table 2.4.

2.4   Proteins from Amino Acids

The cell contains proteins normally about 30 to 70% of its dry weight. The protein typically made up of four elements i.e. carbon, hydrogen, nitrogen and oxygen. The average weight percentages of these elements in proteins are about carbon-50%, hydrogen-7%; oxygen-23%; and nitrogen-16%. Protein also contains about 3% of sulphur which contributes to the three-dimensional stabilization of the protein by forming S-S disulphide bonds. The protein configurations fibrous and globular are shown in the Fig. 2.9. Proteins function as biocatalysts and the biocatalysts are known as enzymes. And enzymes are found in various locations of the cell, for example, in cytoplasm, in membranes or in larger molecular assemblies.

Some proteins contribute to structure of the cell membranes. Some proteins function as flagella in the single cell organisms. Flagella contain contractile proteins which helps the cell in motion. Different techniques can be adopted to characterize, isolate, and purify the proteins.

Building blocks of proteins

Aminoacids are the monomeric building blocks of the polypeptides and proteins. The general structure of the monomeric building block is,

$$H_2N \longrightarrow \overset{\displaystyle H}{\underset{\displaystyle R}{C}} \longrightarrow COOH$$

Amino acids are recognized as per the R group attached to the α – carbon. Most of the organic compounds like sugars and amino acids are optically active, which contain asymmetric carbon. They occur in two isometric forms as shown for amino acids.

Table 2.5 The various biological functions of proteins.

Protein	Function/Occurrence
Enzyme (biological catalysis) :	
Glucose isomerase	Isomerizes glucose to fructose
Trypsin	Hydrolyzes some peptides
Alcohol dehydrogenase	Oxidizes alcohols to aldehydes
RNA polymerase	Catalyzes RNA synthesis
Regulatory proteins :	
Lac repressor	Controls RNA synthesis
Catabolite activator protein	Catabolite repression of RNA synthesis
Interferons	Induce virus resistance
Insulin	Regulates glucose metabolism
Bovine growth hormone	Stimulates growth and lactation
Transport proteins :	
Lactose permease	Transports lactose across cell membrane
Myoglobin	Transports O_2 in muscle
Hemoglobin	Transports O_2 in blood
Protective proteins in vertebrate blood :	
Antibodies	Form complexes with foreign molecules
Thrombin	Component of blood clotting mechanism
Toxins :	
Bacillus thuringiensis toxin	Kills insects
E.coli ST toxin	Causes disease in pigs
Clostridium botulinum toxin	Causes bacterial food poisoning
Sotrage proteins :	
Ovalbumin	Egg-white protein
Casein	Milk protein
Zein	Seed protein of corn
Contractile proteins :	
Dynein	Cilia and flagella
Structural proteins :	
Collagen	Cartilage, tendons
Gylcoproteins	Cell walls and coats
Elastin	Ligaments

Generally only the L-amino acids are found in most proteins and D-amino acids are rarely found in amino acids. In aqueous solutions the acid-COOH and base –NH$_2$ groups of amino acids ionize. At low pH the amino acid is positively charged, at high pH it is negatively charged and at intermediate pH the amino acid acts as a dipolar ion or zwitterion. This pH at which there no net charge is the isoelectric point, and varies according to the R-group involved. At an isoelectric point the amino acid will not migrate under the influence of an applied electric field. At isoelectric point minimum solubility is exhibited by an amino acid. Because of these properties amino acids can be separated from mixtures by using separation techniques like ion exchange, electro-dialysis and electrophoresis.

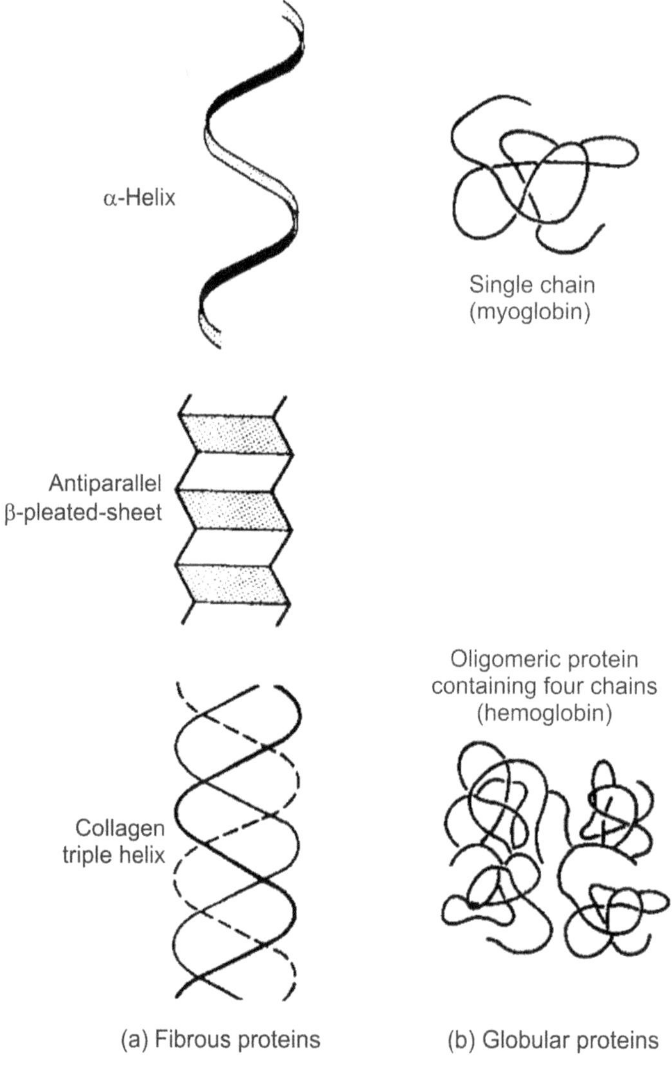

Fig. 2.9 Protein structure types 1. Fibrous 2. globular

The 20 amino acids are commonly found in proteins (Fig. 2.9). Similar to carboxyl, and amino acid groups, some R groups can also ionize. the amino acids that possess no polar R groups are hydrophobic in nature and other R groups are hydrophilic in nature. Simple proteins are the biopolymers produced by the condensation of amino acids. The condensation reaction is between amino group of one amino acid and carboxyl group of other amino acid. Thus a peptide bond is formed between them.

The reaction is given here.

Polypeptides are the result of short condensation chains of amino acids. The examples for polypeptides are, insulin, somatostatin. Larger amino acid chains are known as proteins and contain 50 to 100 amino acid residues. The average molecular weight of each residue is about 120 and the protein molecular weights are about 10, 0000. Total hydrolysis of proteins is carried out by heating them in 6 N HCl for 10 to 24 hours at $100\,^{\circ}C$ to $120\,^{\circ}C$ temperature. Amino acids are recovered in hydrochloric form.

References

1. Michael J.Pelczar, E.C.S.Chan and Noel R Krieg (1996) Microbiology, Tata McGraw-Hill, Publishing Company limited, New Delhi.

2. Baily, J.E. and Ollis, D.F. (1986) Biochemical Engineering Fundamentals, McGraw Hill, Inc., New York.

3. Lehninger, A.L. (1983) Biochemistry, Worth Publishers, Inc., New York.

Review Questions

1. What are the two types of microbial cells?
2. Describe various forms of bacteria.
3. What is DNA?
4. What is RNA?
5. What are proteins?
6. Write about amino acids.
7. Write about protein structure.
8. Write about structure of a cell.

CHAPTER 3

Metabolic Pathways

It is very essential to select an organism which can efficiently generate product in any given process. He genes can be added or removed from the organisms of the metabolic functions depend on the metabolic pathways. Some important metabolic pathways are discussed here.

The microbial metabolism may be due to genetic changes or to their reponses in accordance with the changes in the environment. Different products may be produced by the same microorganism is they are grown in different medium, also in the different environment conditions. The metabolic pathways can be controlled nutritional and environment condition in any bioprocess. Under anaerobic conditions ethanol is produced by *Saccharmyes cerviseae* and other under aerobic condition the major product is yeast cells. However under aerobic condition also ethanol is produced, if high glucose concentrations are used. During catabolism a compound is divided into smaller compounds is glucose to CO_2 and water. And energy is produced during catabolism during more complex compounds is glucose to glycogen and energy is consumed in the process.

3.1   Energy in Metabolic Reactions

Metabolic reactions of the living cells like, biosynthesis nutrient transport, motility and maintenance need energy, for their functions. Catabolism of carbon compounds, like carbohydrate provide energy required for the cell.

Metabolism reactions vary from organism to organism and they are complicated. These reactions are basically categorized into three types (Fig. 3.1).

1. Degradation of nutrients
2. Biosynthesis of small molecules i.e., amino acids, nucleotides,
3. Biosynthesis of large molecules.

These reactions are carried out simultaneously. Metabolic reactions are responsible for the production and release of end products from the cells. The end products include valuable products like antibiotics, organic acids and amino acids.

Biological systems basically store and transfer energy through adenosine triphosphate (ATP). ATP contains high energy phosphate bonds. The standard free energy charge for the ATP hydrolysis is about 7.3 kcal/mol. However the actual free energy release in the cell may be much higher because the concentrations and ATP is higher than for Adenosine diphosphate.

$$ATP + H_2O \Leftrightarrow ADP + P_i \; \Delta G^0 = -7.3 \text{ kcal/mol.} \qquad .....(3.1)$$

Similarly when ATP is formed from ADP energy is stored, and ATP dissociates to ADP and releases energy.

$$ADP + H_2O \Leftrightarrow AMP + P_i \; \Delta G^0 = -7.3 \text{ kcal/mol.} \qquad .....(3.2)$$

Even though analog compounds like –guanosine triphosphate (GTP), Undine Triphosphate (UTP) and Cytidine triphosphate (CTP) store and transfer high energy phosphate bonds they are not equal to ATP.

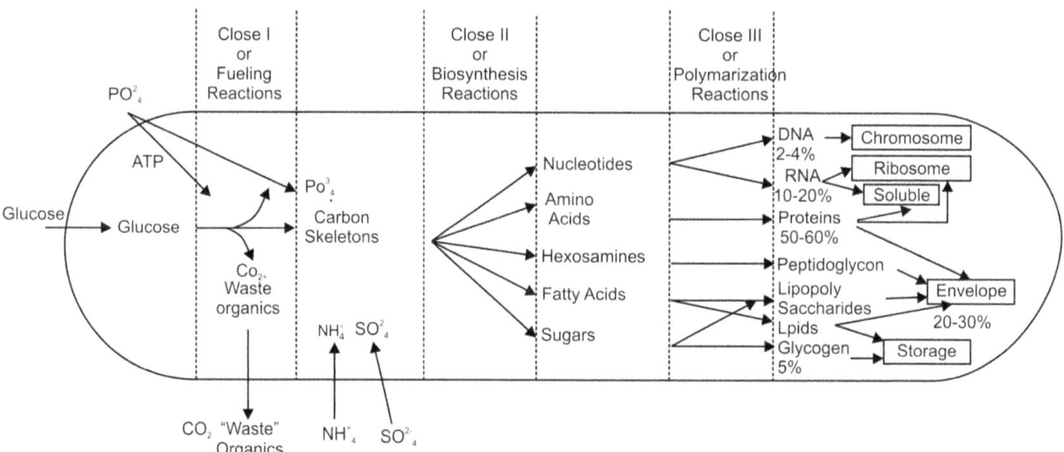

Fig. 3.1 Typical biochemical reactions in a microorganism.

The energy stored in the ATP is given to lower energy compounds like glucose 6-phosphate and glycerol 3-phosphate.

Fig. 3.2 the stored energy in ATP taken for higher energy compounds is given to lower energy compounds.

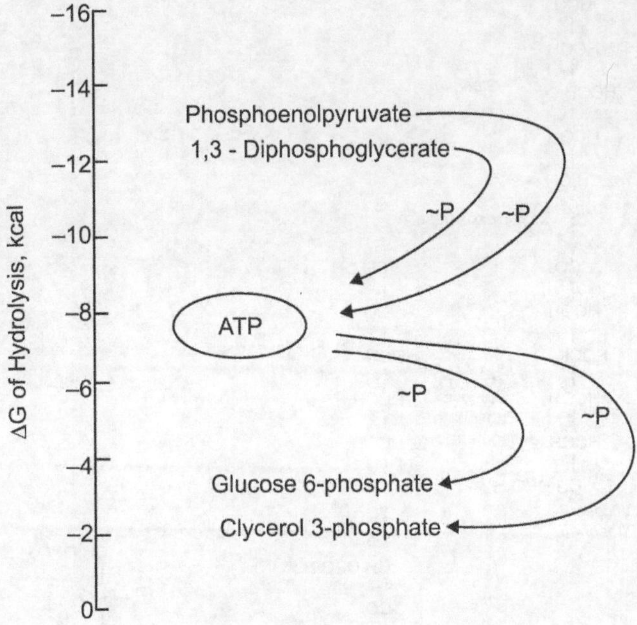

Fig. 3.2 Biological energy transferred from high energy to low energy compounds through Adenosine Triphosphate.

Glucose plays an important role in providing carbon and energy to microorganism. The catabolism of the glucose is carried out by various metabolic pathways in microorganism. The primary step is catabolism of glucose in glycolysis or Embden – Meyerhof – Parnas (EMP) pathway.

The catabolism of glucose to Pyruvate conversion

1. EMP pathways – Glucose to Pyruvate conversion.

2. Kerbs, TCA or citric acid cycle – Pyruvate conversion to CO_2 and NADH.

3. Respiratory or electron transport chain – Transfer of electron from NADH to chain electron accepted during formation of ATP. Aerobic or anaerobic respiration is classified due to the electron acceptor mechanism. In aerobic respiration oxygen is used a final electron acceptor. In anaerobic respiration NO^-_3, SO^{3-}_4, Fe^{3+}, and S^0 are used as electron acceptors.

3.2   Embden - Meyerhof – Parnas (EMP) Pathway

In EMP pathway or glycolysis glucose is broken down to two pyruvate molecules. (Fig. 3.3)

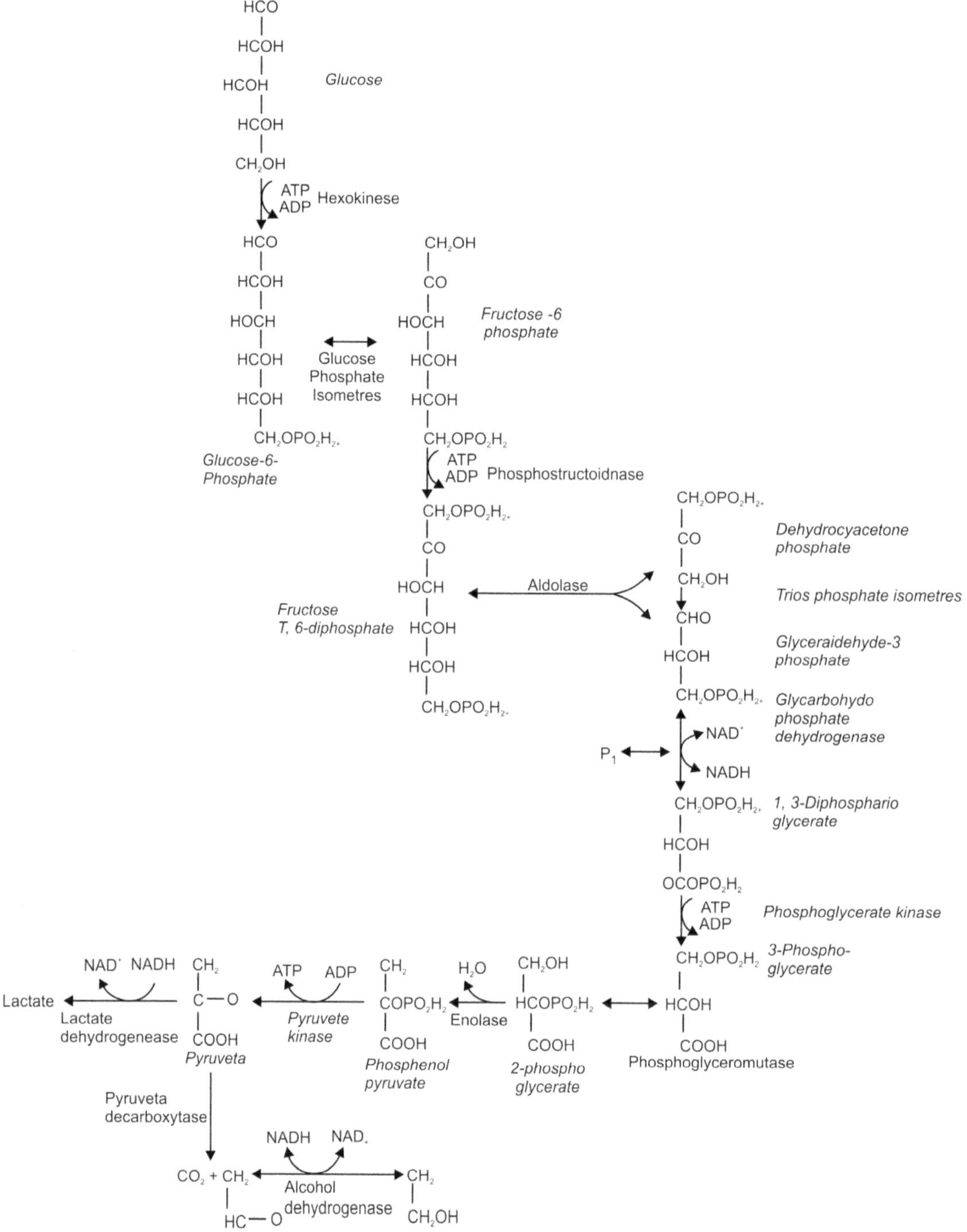

Fig. 3.3 Break down of Glucose by Glycolysis and the reactions involved.

Initially glucose is phosphorylated to glucose 6-phosphate by hexokinase enzyme. Glucose 6-phosphate may be kept inside the cell. In the process the glucose 6-phosphate is converted to fructose 6 phosphate by phosphoglucose isomerase and to fructose 1-6 diphosphate by phospho fructokinase enzyme. The first and third reactions are the only two ATP consuming reactions in EMP pathway, and they are irreversible. In the fourth step, the fructose 1, 6-diphosphate is broken down into dihydroxy acetone phosphate and glyceraldehyde 3-phosphate by aldolase. Dihydroxy acetone phosphate and glyceraldhyde 3 phosphate are in equilibrium. Glyceraldehyde 3 phosphate is used in glycolysis and glyceraldehydes 3 phosphate is converted into glyceraldehydes 3 phosphate. Glyceraldehyde 3-phosphate is first oxidized with the addition of inorganic phosphate (Pi) into 1, 3 diphosphoglycerate by glyceraldehyde 3-phosphate dehydrogenase enzyme. 1, 3-diphosphoglycerate releases one phosphate group to from ATP from ADP and then is converted into 3 phosphate by 3 phosphoglycerate kinase. In 8th step 3-phosphoglycerate is converted into 2-phosphoglycerate by phosphoglyceromutase enzyme. In 9th step 2-phosphoglycerate is dehydrated in to phosphenol Pyruvate by enolase. In 10th step phosphoenol pyruvate is further dephoshorylated to pyruvate by pyruvate kinase enzyme and ATP is also formed. The end product pyruvate is an important metabolite in metabolism. During anaerobic conditions, pyruvate may be converted into lactic acid, ethanol or other products such as acetone, butanol and acetic acid. Glucose conversion into products like ethanol, butanol, etc, anaerobically is often known as fermentation.

3.3   Tri Carboxylic Acid - Cycle

In TCA cycle pyruvate is converted into CO_2 and NADH under aerobic conditions. The pyruvate produced in EMP Pathway; transfer its reduction power to NAD^+ through TCA cycle. While glycolysis occurs in cytoplasm, the TCA cycle occurs in matrix of mitochondria in eukaryote cells. In prokaryotes these reactions are happening with membrane bond enzymes.

Enter into TCA cycle is provided by the acylation of coenzyme A by pruuvate.

$$\text{Pyruvate} + NAD^+ + \text{Co- A-SH} \xrightarrow{\text{Pyruvatedehydrogenase}} \text{acetyl COA} + CO_2 + NADH + H^+.$$

$$\dots\dots(3.3)$$

Acetyl COA is transferred by the conversion of the NADHs produced in glycolysis to 2 FADH through mitochondrial membrane.

Acetyl COA is the important intermediate in amino acids and fatty acids metabolism.

The reactions involved in TCA cycle are shown in Fig. 3.4.

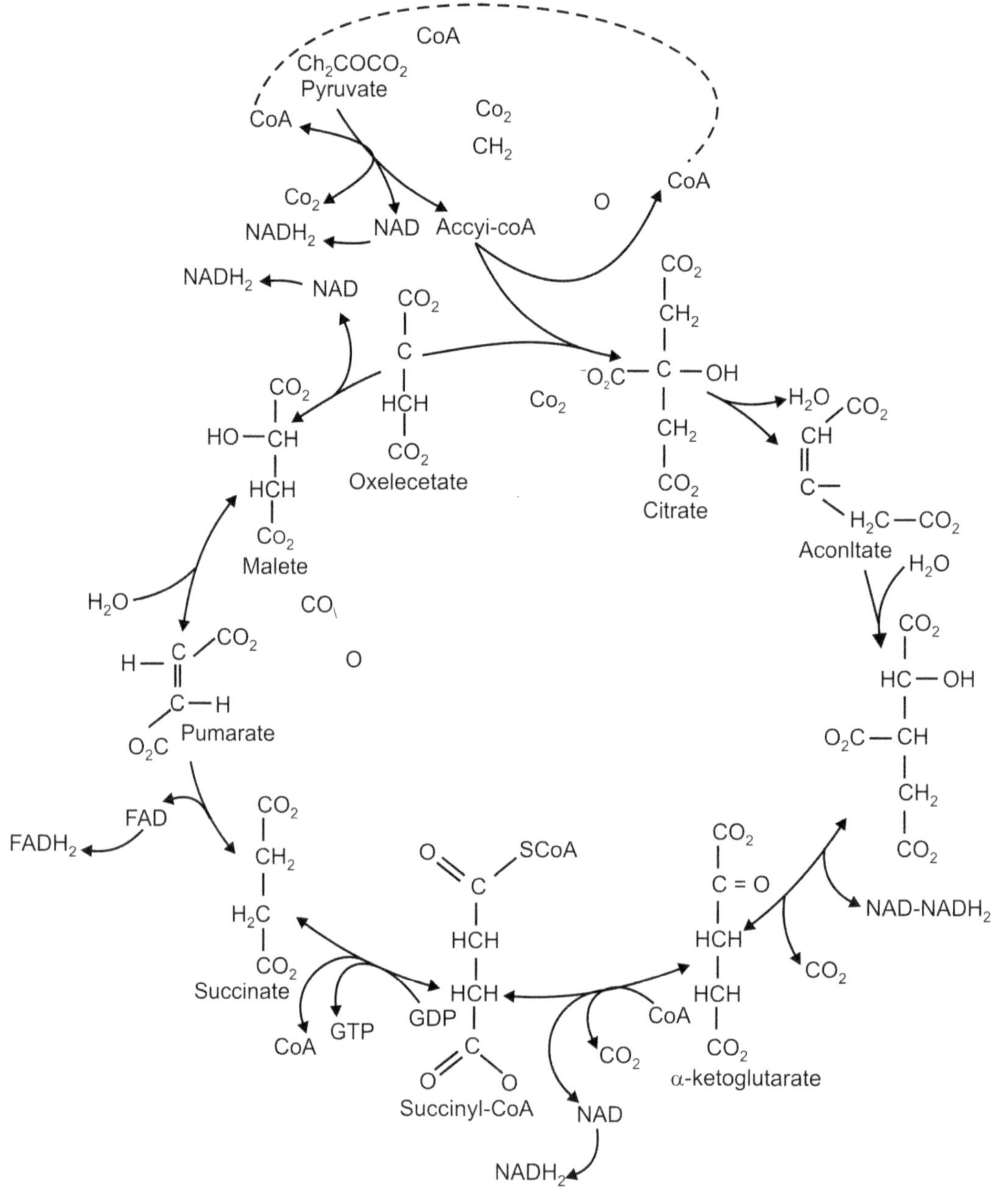

Over all reaction :   Pyruvate + 4NAD + FAD $\rightarrow$ 3CO₂ + 4NADH₂ + FADH₂
GDP + phosphorous $\rightarrow$ GTP
GTP + ADP $\rightarrow$ GDP + ATP
Oxidative phosphorylection : 4NADH₂ ≡ 12ATP — 15 ATP
FADH₂ ≡ 2ATP

Fig. 3.4 TCA cycle with energy yields.

TCA cycle as seen in eukaryote cells (Fig. 3.4).

Overall stoichiometry plus reaction:

$$\text{Pyruvate} + 4\ NAD + FAD \rightarrow 3CO_2 + 4\ NADH_2 + FADH_2 \qquad \ldots\ldots(3.4)$$

$$GDP + \text{Phosphate} \rightarrow GTP$$

$$GTP + ADP \rightarrow GDP + ATP$$

Acetyl CoA condenses with oxaloacetic acid to form citric acid , which is again converted to isocitric acid and then to α- Keto glutaric acid is decaroxylated and oxidized into succinic acid, which is again oxidized into fumaric acid . Fumaric acid is hydrated to malic acid and malic acid is oxidysed to oxaloacetic acid . These are the two last steps of TCA cycle. In TCA cycle for each pyruvate molecule entering cycle, three CO_2 four $NADH^+\ H^+$ and one $FADH_2$ also produced. The α- Keto glutarate and succinate produced in the TCA cycle are used as precursors for the synthesis of some amino acids. So the reducing Power ($NADH^+$, H, $FADH^2$) produced is used either biosynthetic pathways or for ATP generation through the electron transport chain.

The TCA cycle reaction is given below

$$\text{Acetly}\ \ CoA + 3\ NAD^+ + FAD + GDP + P_i + 2\ H_20\ \rightarrow CoA + 3(\ NADH + H^+\ ) + FADH_2$$
$$+\ GTP + 2\ CO_2 \qquad\qquad \ldots\ldots(3.5)$$

The GTP can easily be converted into ATP

The importance of TCA cycle are

(a) Providing electrons (NADH) for the electron transport chain and biosynthesis

(b) Supplying carbon 'C' for amino acid synthesis and

(c) Generating energy

The intermediates in TCA cycle used in biosynthesis extensively the functioning of TCA cycle can be maintained by fixing CO_2 (heterotrophic CO_2 fixation) by cell.

Pyruvate mixed with CO_2 by using the reducing power of one molecule of NADPH+ H + and by Catalytic activity of malic enzyme. Culturing of microorganism may be carried out by fixing heterotrophic CO_2 .

3.4   Electron Transport Chain

The sequence in the respiration chain is commonly known as electron transport chain. The formation of ATP from the transport chain is called as oxidative phosphyorylation. The electron that are carried out by $NADH + H^+$ and $FADH_2$ are transferred to oxygen by series of electron carries and ATPs are thus formed. In eukaryotes each $NADH^+$, H^+, produces three ATPs and each $FADH_2$ two ATPs. The electron transport chain is given in Fig. 3.5 Oxidation of NADH and the flow of electrons transport system resulting in transfer of protons (H^+) from inside to outside of membrane is shown. The tendency of protons to return to the inside is known as proton motive force.

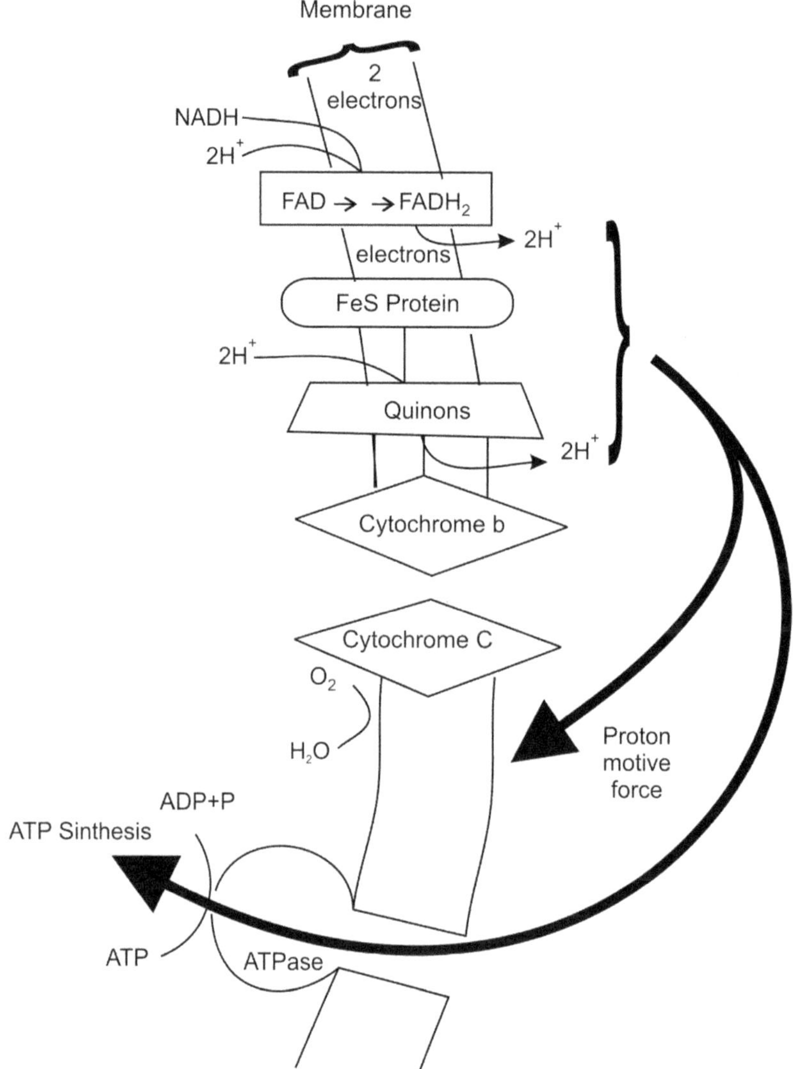

Fig. 3.5 Electron transport and electron transport phosphyorylation.

In the bottom of the Fig. 3.5 it is shown that ATP synthesis occurs as protons enter the cell. An ATPase are enzymes uses the proton – motive force for synthesis ATP.

The significance of the electron transport chain to generate NADS for glycolysis and ATPs for biosynthesis.

The P/0 ratio is used to indicate the number of phosphate bonds made for each oxygen atom used as an electron acceptor.

Ex. $\frac{1}{2} O_2 + NADH + H^+ \rightarrow H_2O + NAD^+$

By Aerobic catabolism of glucose $NADH^+$, H^+, $FADH_2$ ATP are formed and these are tabulated in Table 3.1.

Table 3.1 Summary of NADH. FADH$_2$, and ATP Formation during Aerobic Catabolism of Glucose (Based on Consumption of 1 Mole of Glucose).

	NADH	FADH$_2$	ATP	Total ATP[a]
Glycolysis	2	-	2	6[a]
Oxidative decarboxylation of pyruvate	2	-	-	6[b]
TCA cycle	6	2	2	24
Total	10	2	4	36 mol ATP

a – P/O ratio is 3 for NADH and P/O ratio 2 for each FADH$_2$

b- Yield is 6 ATP instead of 8 because 2 NADH are converted to 2 FADH$_2$ for the transfer to acetyl CoA into mitochondria.

P/O ratio 3 implies 3 ATP phosphates are made for each proton transported by the electron transport chain.

The overall reaction when 3 ATP produces for each NADH of aerobic glucose catabolism in eukaryotic cells is.

$$\text{Glucose} + 36\,P_i + 36\,ADP + 6O_2 \rightarrow 6CO_2 + 6\,H_2O + 36\,ATP \qquad .....(3.6)$$

The energy deposited in 36 mole of ATP is ($7.3 \times 36 = 262.8$) 263 Kcal/mol glucose. The free energy change in the direct oxidation of glucose is 686 kcal/mol glucose. Hence the energy efficiency, of glycolysis in 38% during normal conditions, by the correcting factor for non standard condition this efficiency would be more than 60%. The remaining energy stored in glucose molecule will be dissipated as heat. In prokaryotic cells the conversion of the reducing power to ATP is less efficient. So the number of ATP generated NADH$^+$, H$^+$, is normally ≤ 2 and only one ATP may be produced for each FADH$_2$. Finally in procarytic cells is single glucose molecule will yield less than 24 ATP and P/O ration lies 1 and 2.

3.5   Aerobic Glucose Metabolism

In glycolysis most of the enzymes or regulated by feedback inhibition.

Phosphyorylation of fructose 6 phosphate by phosphofructo kinase is the important control site in glycolysis.

$$\text{Frutose 6 phosphate} + 6\,ATP \rightarrow \text{fructose 1,6 diphophate} + ADP \qquad .....(3.7)$$

The enzyme phosphofructo kinase is an allosteric enzyme which is activated by ADP + Pi but inactivated by ATP.

$$\text{Phosphofructo kinase (active)} \xleftarrow[ADP]{ATP} \text{Phosphofructo kinase (inactive)} .....(3.8)$$

When the ATP/ADP ratio is high, the enzyme is inactivated and glycolysis rate is reduced, Also the ATP synthesis is reduced.

The rate of glycolysis under anaerobic conditions will be higher than that under aerobic conditions. In the presence of oxygen ATP yield is high due to TCA and electron transport chains functioning.

Hence due to high levels of ATP, ADP and Pi becomes limiting resulting in inhibition of Phosphofructo kinase. Some enzymes of glycolysis with –SH groups are inhibited by high levels of oxygen. High ratio of NADH/NAD$^+$ also reduces the rate of glycolysis.

Some enzymes of Krebs cycle are regulated by feed back inhibition. For example Pyruvate dehydrogenase is inhibited by ATP, NADH and acetyl CoA and activated by ADP, AMP and NAD$^+$. Also citrate synthetase is inactivated by ATP and activated by ADP and AMP Succinyl CoA synthetase is inhibited by NAD$^+$. It may be observed that high ATP/ADP and NADH/NAD$^+$ ratios reduce, the rate of TCA cycle.

Some steps in electron transport chain could be inhibited by cyanide azide, carbon monoxide and antibiotics.

Metabolism of Hydrocarbons

Aliphatic and aromatic hydrocarbons metabolism play an important role in bioprocesses.

Aliphatic hydrocarbons:

Metabolism in general requires oxygen and only few organism like Pseudomonas, Mycobacteria, Saccharomyces sp. and molds can metabolize hydrocarbons. The low solubility of hydrocarbon in water is a constraint for rapid metabolism.

The first step in metabolism of aliphatic hydrocarbons is oxygenation by oxygenases. Hydrocarbon molecules are converted to an alcohol by addition of oxygen into the end of carbon chain. Also the alcohol molecule is further oxidized to aldehyde and then to organic acid. This is further converted into acetyl coA, which is metabolized by TCA cycle.

Aromatic hydrocarbons:

Aromatic hydrocarbons are subjected to oxidation by this action of oxygenases and this reaction is slower than those of aliphatic hydrocarbons. The important intermediate in this oxidation is catchol and it is further broken down to acetyl-CoA or TCA cycle intermediate.

The aerobic metabolism of benzene is given below

$$\text{Benzene} \rightarrow \text{catechol} \rightarrow \text{cis-cis muconate} \rightarrow \beta \text{ Keto adipate} \rightarrow \text{acetyl CoA } + \text{ succinate}$$

$$.....(3.9)$$

Anaerobic metabolism of hydrocarbons is very much difficult, only a few microorganisms can metabolic hydrocarbons under anaerobic conditions.

3.6   Biosynthesis

Precursors required for the biosynthesis of amino acids, nucleic acids, lipids, and polysaccharides are provided by the TCA cycle and glycolysis. Important pathways are desired here.

The first pathway is Pentose – Phosphate or hexose mono phosphate pathway Fig. 3.6. This pathway generates significant reducing power which can be used to supply energy to the cell. Its main role is to provide carbon skeletons for biosynthesis reactions and also reducing power necessary to support anabolism. It is known that NADPH is used in biosynthesis while NADH is used in energy production. This pathway provides a list of small organic compounds with three, four, five and seven carbon atoms. The compounds thus produced play an important role for the synthesis of ribose, purines, coenzymes and aromatic amino acids. Glyceraldehyde 3 phosphate produced will be oxidized to yield energy through conversion to Pyruvate and further oxidation of Pyruvate in the TCA cycle.

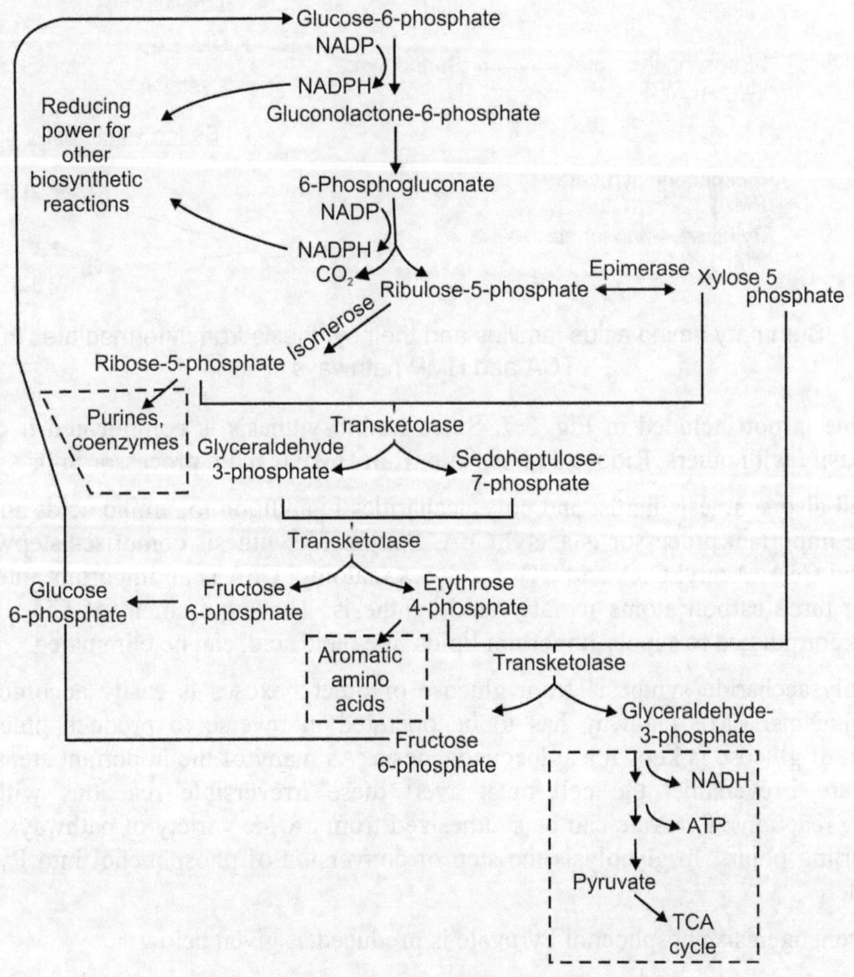

Fig. 3.6 Pentose phosphate pathway.

The production of amino acids consumes a large amount of cellular building block. The 20 amino acids are classified into various families. Summarized of amino acids further are given in Fig. 3.7.

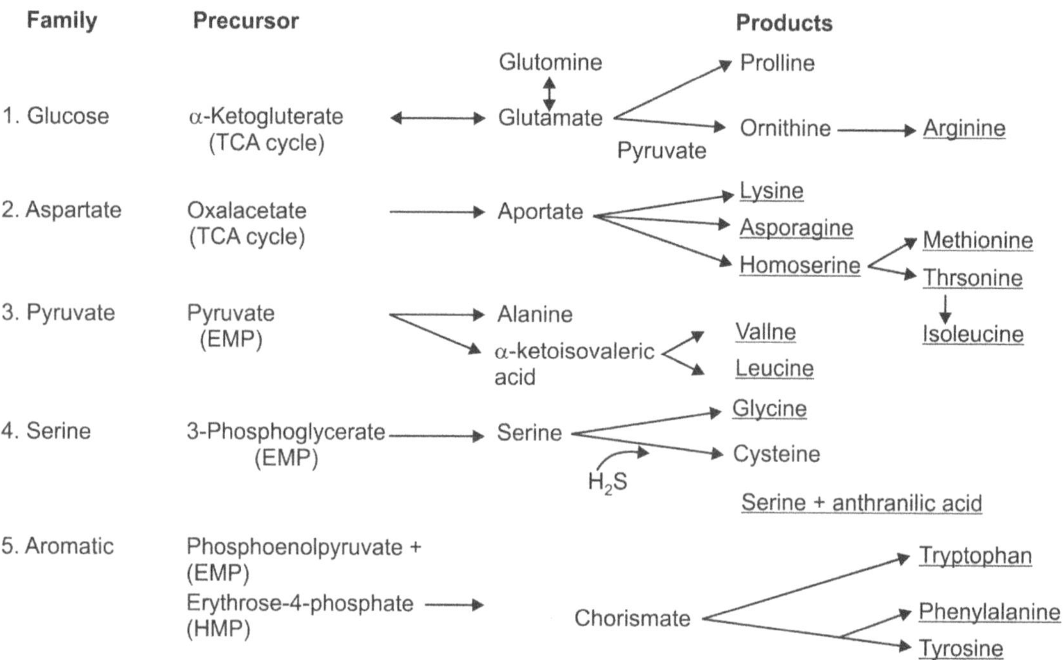

Fig. 3.7 Summary amino acids families and their synthesis from intermediates in EMP, TCA and HMP pathways.

Histidine is not included in Fig. 3.7. Since its biosynthesis is complicated it cannot be grouped easily with others. Ribose 5 phosphate from HMP is main processor in its synthesis.

The cell also synthesis lipids, and polysaccharides in addition to, amino acids and nucleic acids. The important processor is acetyl CoA. Fatty acid synthesis comprises stepwise build up of acetyl CoA. Acetyl CoA and CO_2 produces malomyl CoA is an important intermediate containing three carbon atoms in fatty acid synthesis. The requirement of CO_2, when the medium is formulated to supply important lipids like oleic acid, can be eliminated.

The polysaccharide synthesis from glucose or other hexoses is easily accomplished in many organisms. EMP pathway has to be operated in reverse to produce glucose. The production of glucose is known as gluconeogenesis. As many of the important steps in EMP pathway are irreversible, the cell must avert these irreversible reactions with energy consuming reactions. Pyruvate can be synthesized from a wide variety of pathways so it will be the starting point. In glycolysis the step of conversion of phosphoenol into Pyruvate is irreversible.

In gluconeogenisis phosphoenol Pyruvate is produced as given below

$$\text{Pyruvate} + CO_2 + \text{ATP} + H_2O \rightarrow \text{Oxalacetate} + \text{ADP} + P_i + 2H^+ \qquad .....(3.10)$$

$$\text{Oxalacetate} + \text{ATP} \rightarrow \text{Phosphoenol} + \text{ADP} + CO_2 \qquad \text{.....(3.11)}$$
$$\text{Pyruvate}$$

Net reaction :

$$\text{Pyruvate} + 2\,\text{ATP} + H_2O \rightarrow \text{Phosphoenol pyruvate} + 2\,\text{ADP} + 2\,H^+ \qquad \text{.....(3.12)}$$

The glycolysis reactions , up to formation of fructose 1, 6 diphosphate are reversible under appropriate conditions, gluconeogenesis can be completed with, fructose 1,6 diphosphate, glucose 6 phosphate and these are not present in EMP.

References

1. Baily, J.E. and Ollis, D.F. (1986) Biochemical Engineering Fundamentals, McGraw Hill, Inc., New York.

2. Black, J.G. Microbiology, Principles & Applications 3 rd Ed. Prentice Hall, Upper saddle, river, NJ, 1996.

3. Neway J.O. Fermentation Process Development of Industrial Organisms. Marcel-Dekker, Inc N.Y. 1989

4. Michael L. Shuler, Fikret Kargi. Bio Process engineering Basic concepts 2 nd edition, Prentice Hall of India, New Delhi – 2002.

Review Questions

1. Write about typical biochemical reactions in microorganism.
2. Explain TCA cycle.
3. What is glycolysis explain ?
4. Write about aerobic metabolism.
5. Explain pentose phosphate pathways.

CHAPTER 4

Microbial Growth

Microorganism require desirable physical chemical and nutritional conditions for their growth. They grow in number as well as in their size. The microorganism take nutrients from the medium and they use it for the energy production. Partly it is used for the biosynthesis and formation of products.

$$\text{microbial cells} + \text{substrate} \rightarrow \text{more of microbial cells} + \text{products (extracellular)}$$

$$X + \Sigma S \rightarrow nX + \Sigma P \qquad \qquad \dots\dots(1)$$

The growth rate is proportional to the concentration of the cell. Specific growth rate depicts, rate of microbial growth and is represented by

$$\mu_{net} X = \frac{dx}{dt} \qquad \qquad \dots\dots(2)$$

Where

X = cell mass concentration (g/l)

t = time (hr)

μ_{net} = specific growth rate (hr^{-1})

During the growth of microbial cells, they consume material from cells environment and convert them to metabolic end products.

Unlike the chemical reactions which depends on only reactants, microbial reactions depend on various like type of cell, their growth etc.

The microbial cells or their unicellular organisms or molds affect the cell growth and physical and chemical characteristics of reacting medium (Fermentation broth). During the process of microbial growth, nutrients and substrate are consumed simultaneously to convert them into metabolic products and on the different ages of the microbial cells.

4.1 Microbial Cell Growth Phases

After inoculation, microorganism grow in the liquid medium, by consuming the nutrients present in the medium. The growth cure consists of following phases

(a) Lag phase

(b) Logarithmic (or) exponential phase

(c) Stationary phase

(d) Death phase

Lag phase

The lag phase starts immediately after inoculating the microbial cells to the growth medium. At this time the microbial cells adjusts themselves to the new environment. Their molecular constituents are reorganized during their transfer to the new medium. The microbial cells synthesize new enzymes suiting to the new-medium while the synthesis of some other enzymes is repressed. During this time the cell mass may be increased and there will not be any change in the number of cells.

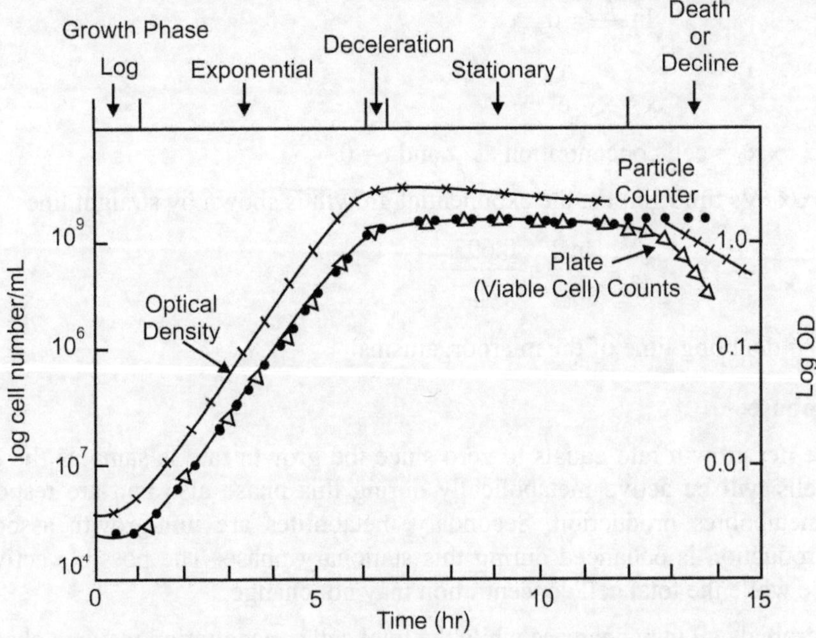

Fig. 4.1 Growth curve for microorganism (bacteria).

The long lag phase may be caused by the low concentration of nutrients or growth factors.

The Inoculum cultures age plays an important role on the lag phase period. Normally the lag phase is proportional to the Inoculum age. The lag phase period may be by adopting the microbial cells to the growth medium prior inoculation. Preferably the microbial cells must be young and active. Also the inoculum size may be sufficient large i.e 5 – 10% by volume. Some growth factors medium be added and optimization of nutrient medium may be adopted to reduce the lag phase.

Exponential growth phase

During this phase microbial cells which are already adapted to the new environment, start multiplying rapidly. The cells multiply rapidly resulting in increase of cell mass as well as cell number. Interestingly most of the components of the cell grow at the same rate during this phase. Also the net specific growth rate based on cell number or cell mass may be equal.

The growth rate is not dependent upon the concentration of nutrients since their concentration is large. The growth rate of organism during this phase follows first order.

$$\frac{dx}{dt} = \mu_{net} X \qquad \qquad .....(3)$$

$$x = x_0 \quad \text{at} \quad t = 0$$

upon integration we get ,

$$\ln \frac{x}{x_0} = \mu_{net} t \qquad \qquad .....(4)$$

or $\qquad \qquad x = x_0 \ e^{\mu_{net} t}$

Where $\qquad x, x_0 = $ cell concentration at t and t = 0

In the ln x/x_0 Vs time graphs, the exponential growth is shown by straight line

$$\tau_d = \frac{\ln 2}{\mu_{net}} = \frac{0.693}{\mu_{net}} \qquad \qquad .....(5)$$

Where $\tau_d = $ doubling time of the microorganisms

Stationary phase

In this phase net growth rate equals to zero since the growth rate is same as the death rate. Microbial cells will be active metabolically during this phase also and are responsible for secondary metabolites production. Secondary metabolites are non-growth associated and metabolic production is enhanced during this stationary phase. The possible activities cells may decrease while the total cell concentration may not change.

Viable microbial cell may decrease while the total cell concentration may not change.

Viable cells may be reduced due to cell lysis. Also some microbial cells may grow on lysis products of lysed cells. Secondary metabolites may be produced and microbial cells may not be growing.

Endogenous metabolism namely catabolyzing of cell reserves for new building blocks and for energy producing monomers, takes place during the stationary phase.

The energy is also utilized by the cell to maintain motility and repair of damaged structures as a part of metabolic function. The conversion of cell mass into maintenance energy, during stationary phase, can be described by the following equation

$$\frac{dx}{dt} = -k_d x \qquad \qquad .....(6)$$

Or $\qquad \qquad X = X_0 \ e^{-kdt}$

Where

$\qquad K_d = $ Decay constant or first order rate constant for endogenous metabolism

$\qquad X_0 = $ Cell mass concentration at the onset of stationary phase.

The growth of cells may be reduced either due to depletion of essential nutrients or accumulation of toxic compounds or products. The product, which is produced or accumulated in the medium, may inhibit the growth and the inhibitor may stop the growth of cells at an increase to certain level. For example, the ethanol produced during sugar fermentation will inhibit the growth of yeast *(Saccharomyes cerviseae)*

Death phase or Decay phase

Even though the decay of cells may be started during stationary phase, the distribution between stationary and decay phase is not always possible. The death or decay phase starts at the end of stationary phase, either due to accumulation of toxic compounds or due to depletion of nutrients which are required for the growth of cells.

The death or decay rate follows first order kinetics and is represented by the following equations.

$$\frac{dN}{dt} = -K'_d N \qquad \qquad .....(7)$$

or

$$N = N_s e^{-k'_d t}$$

Where　N_s = concentration of cells at the end of stationary phase

K_d = first order death rate constant

It is interesting to note that during the onset of death phase, if nutrient rich medium is provided, the same cells may grow again. The cell will die if it is not replenished with nutrient rich medium.

4.2　Yield Coefficient

Yield coefficient are defined based on consumption of a material

$$Y_{x/s} = \frac{-\Delta X}{\Delta S} \qquad \qquad .....(8)$$

The pattern of substrate utilization in a cell can be expressed as given below

$$\Delta S = \Delta S \quad + \quad \Delta S \quad + \quad \Delta S \quad + \quad \Delta S \quad .....(9)$$

$$\text{(Assimilation} \quad \text{(assimilated into an} \quad \text{(growth} \quad \text{(Maintenance}$$
$$\text{into biomass)} \quad \text{extracelluar product)} \quad \text{energy)} \quad \text{energy)}$$

Yield coefficient based on substrates and products formed.

$$Y_{x/0_2} = -\frac{\Delta X}{\Delta O_2} \qquad \qquad .....(10)$$

$$Y_{P/s} = -\frac{\Delta P}{\Delta S} \qquad \qquad .....(11)$$

Some typical values of yield coefficient are given below

Table 4.1 Partial list of yield factors of some microorganism.

S. No.	Organism	Substrate	$Y_{x/s}$ (g/g)	$Y_{x/02}$ (g/g)
1.	Candida	Glucose	0.51	1.32
2.	Penicllium chrysogemum	Glucose	0.43	1.35
3.	Saccharomyces cerviase	Glucose	0.50	0.70
4.	Candida utilis	Acetate	0.36	0.70
5.	Pseudomonoas sp.	Methanol	0.41	0.44
6.	Pseudomonas sp	Methane	0.80	0.20

$Y_{x/o2}$ = Yield factor gram of cell formed per gram of O_2 formed

Specific rate of substrate uptake for cellular maintenance is described by maintenance coefficient

$$m = -\frac{[ds/dt]_m}{x} \qquad \qquad .....(12)$$

Endogenous metabolism of biomass components will be used for maintenance energy during stationary phase since little substrate is available. Microbial products can be basically categorized into three major types (Fig. 4.2).

4.3   Growth of Filamentous Organism

A. *Growth Associated Products:*

The products are formed simultaneously along with microbial growth the specific rate of product formation is directly proportional to specific growth rate. μ_g

μ_g (specific growth rate) differ from μ_{net} (net specific growth rate) when endogenous metabolism is non zero (>0).

$$q_p = \frac{1}{X}\frac{dP}{dt} = Y_{P/X}\mu_g \qquad \qquad .....(13)$$

B. *Non-growth associated products:*

The product formed during stationary phase when there is no growth rate. At time the specific rate of product formation remain constant.

$$q_P = \beta = \text{constant} \qquad \qquad .....(14)$$

C. *Mixed growth associated products:*

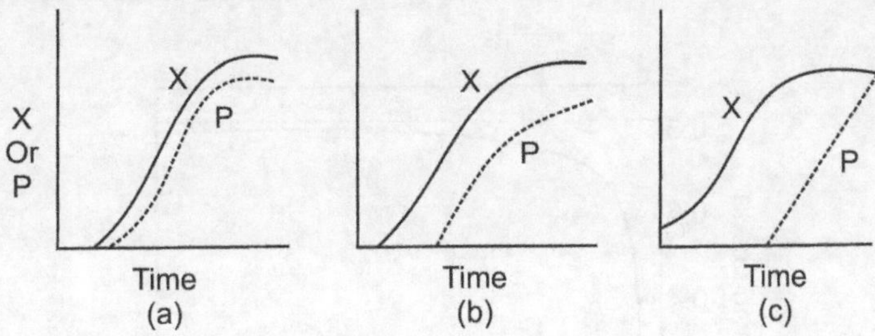

Fig. 4.2 Batch fermentations – Kinetics patterns of Growth and Product formation
(a) Growth associated product formation (b) Mixed growth associated product formation
(c) Non – growth associated product formation.

These products are formed during slow growth and stationary phase. Here the specific rate of product formation is given by,

$$q_P = \alpha\,\mu\,g + \beta \qquad \qquad .....(15)$$

Examples are lactic acid fermentation and Xanthan gum production.

Growth and product formation patterns in fermentations.

1. Growth associated product formation
2. Mixed growth associated product formation
3. Non-growth associated product formation

Luedking – Piret equation

If $\alpha = 0$, the product is only non-growth associated and if $\beta = 0$, the product would be only growth associated and consequently α would be equal to $Y_{p/x}$.

4.4   Growth Kinetics

The growth of microbial cultures at various stages depends upon the nutrient and also on the inhibition. Various models are used to describe the growth kinetic depending on the available environment to the cells. Models can be : 1. structured and segregated, 2. structured and non-segregated and 3. unstructured and segregated and 4. unstructured and non-segregated. Out of all the models, structured and segregated models are realistic but these models are complex in nature.

Unstructured and non-segregated models are described here.

Unstructured and non-segregated models:

Substrate limited growth

The specific growth rate and substrate concentration are related to each other as given in the Fig. 4.3.

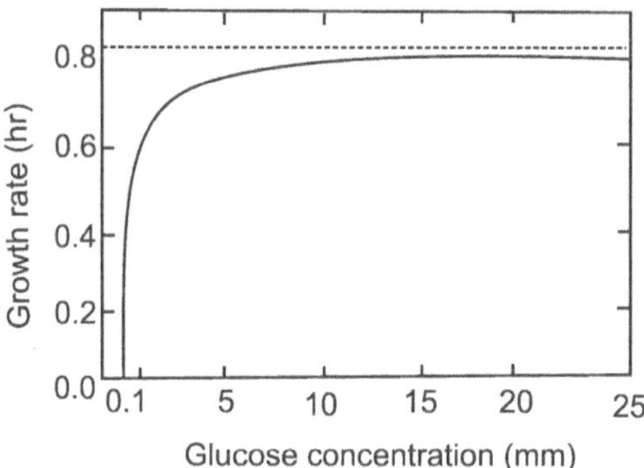

Fig. 4.3 Effect of substrate concentration on specific growth rate.

The growth kinetics can be depicted by Monod equation.

$$\mu_g = \frac{\mu_m S}{K_s + S} \qquad \qquad .....(16)$$

Where

$$\mu_m = \text{Maximum specific growth]}$$

when $\qquad s >> K_s$

if endogenous metabolism is neglected

Then $\qquad \mu_g = \mu_{net}$

K_s is known as saturation constant and it is equal to the concentration of rate limiting substrate when the specific growth rate is equal to the equal one half of the maximum specific growth rate so $K_s = S$ when $\mu_g > \dfrac{\mu_{max}}{2}$.

Monod equation is often used for many systems satisfactorily and it is the unstructured and non-segregated model. Particularly when the cell growth is slow and population density is low, the Monod equation describes substrate limited growth only. The toxic waste products will be released when the consumption of a carbon-energy substrate is rapid. Also the build up of toxic metabolic by-products becomes very important, at high population levels. So for rapidly growing dense cultures, the following rate expressions may be used.

$$\mu_g = \frac{\mu_m S}{K_{s0} + S} \qquad \qquad .....(17)$$

$$\mu_g = \frac{\mu_m S}{K_{s1} + K_{s0}s_0 + S} \qquad \qquad .....(18)$$

Where s_0 = initial concentration of substrate

K_{s0} = dimensional less

Monod equation is often used for many systems satisfactorily and it is the unstructured and non-segregated model. Equation for competitive substrate inhibition.

$$\mu_g = \frac{\mu_m S}{K_s(1+s/K_1)+S} \qquad \dots\dots(19)$$

Substrate inhibition may be alleviated by slow and intermittent addition of substrate to the growth medium.

Production Inhibition

Microbial growth may be inhibited by high concentrations of product. Product inhibition may be competitive or non-competitive.

The equation for product inhibition is given here.

$$\text{Competitive product } \mu_g = \frac{\mu_m S}{K_s(1+p/K_p)+S} \qquad \dots\dots(20)$$

$$\text{Non-competitive product } \mu_g = \frac{\mu_m}{(1+K_s/S)+(1+P/K_p)} \qquad \dots\dots(21)$$

The example for non-competitive product inhibition is fermentation of glucose by yeast for ethanol production. The product ethanol at more than 5% inhibits the growth yeast.

Toxic compounds inhibition

The equation used for descriptive of toxic compounds inhibition are given here.

$$\text{Competitive inhibition} = \mu_g = \frac{\mu_m S}{K_s(1+I/K_I)+S} \qquad \dots\dots(22)$$

$$\text{Non-competitive inhibition } \mu_g = \frac{\mu_m}{(1+K_s/S)(1+I/K_I)} \qquad \dots\dots(23)$$

The presence of toxic compounds in medium may result in the cell death

The equation for such condition is given here.

$$\mu_g = \frac{\mu_m S}{K_s+s} - k_d' \qquad \dots\dots(24)$$

Where k_d' = decay rate constant (hr^{-1})

References

1. Aiba, S., A.E. Humphrey, and N.F. Millis, *Biochemical Engineering*, 2nd ed, Academic Press New York, 1973

2. Bailey, J.E, Mathematical Modeling and Analysis in Biochemical engineering ; Past Accomplishments and Future Opportunities, *Biotechnol* Prog , 14:8, 1998,

3. Bailey, J.E., and D.F. Ollis, *Biochemical Engineering* Marcel Dekker, Inc., New York 1996.

4. Garden, E.L., J$_{R.}$, Fermentation Kinetics and Productivity , Chem ind. Rev (London), 1955, p 154

5. Shuler, M.L., On the Use of Chemically Structured Models for Bioreactors, Chem., Engg, Communications. 36: 161-189, 1985.

Review Questions

1. What is specific growth rate?

2. Describe the growth curve of microorganisms.

3. What is yield coefficient?

4. Explain growth of filamentous organisms.

5. What Luedking –Piret equation?

6. Write about substrate imitated growth.

7. What is product inhibition?

8. Write about toxic compounds inhibition.

CHAPTER 5

Immobilization

5.1 Enzyme Immobilization

The restriction or localization of enzymes on some inert supports can be known as immobilization. Various types of inert supports are used depending on the application.

The advantage of immobilization is that the enzymes can be retained in the reactor in case of free enzyme in solution, the enzyme may escape along with the product. This always leads to the addition of enzymes to compensate the enzymes which is lost along with the product. Also the product will not be pure due to the presence of enzymes.

So immobilization of enzymes have many advantages the immobilized enzymes can be used in continuous mode operations. Product purity is improved and effluent problems are drastically reduced. Enzyme stability will also be improved to a great extent.

5.2 Various Method of Immobilization

The methods of immobilization can be basically classified into two types i.e., entrapment and binding namely surface Immobilization.

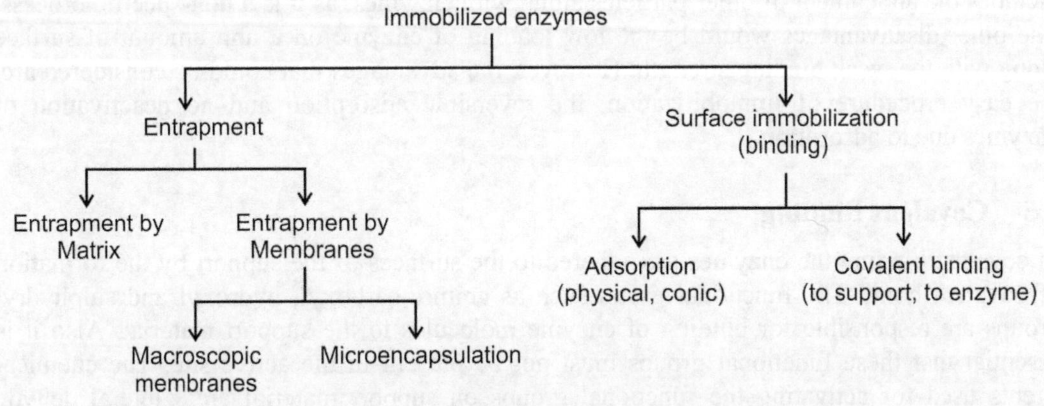

Fig. 5.1 Immobilization Methods.

5.3   Entrapment

The physical process, of immobilization of enzymes is known as entrapment. The two major methods of entrapment are matrix entrapment and membrane entrapment. Polymeric substances like polyacramide gel, calcium alginate and agar are usually used. The advantages of matrix entrapment are that the enzymes are not chemically modified and enzymes properties are retained. The disadvantages may include deactivation of enzyme due to gel formation, enzyme leakage depending on the pore size of the gel and reduced accessibility for substrate due to differential limitations. However enzyme leakage can be avoided by reducing MW (molecular weight) cut off membranes or by reducing the pore size in the solid matrices. Also the diffusional limitations may be averted by reducing the particle size of matrices or capsules. In micro-encapsulation the enzyme is entrapped in microscopic hollow sphere with in porous membrane. Enzymes leakage may also take place in micro-encapsulation. However the advantages of this method is that the enzyme will be good contact with the substrate.

In membrane entrapment, hollow fibers are used to entrap enzyme solution. The membranes commonly used are polyacrylate, cellulose, nylon and poly sulfone. The semi permeable membranes, in all cases, retain high molecular weight compounds while allowing small molecular weight compounds like substrate and products.

Surface Immobilization – adsorption ; covalent bonding:

The surface immobilization comprises adsorption and covalent building.

5.4   Adsorption

In the adsorption the enzymes are physically adsorbed by weak physical forces namely Vanderwaal's forces or dispersion forces. Various types of adsorbents are used but the most frequently used adsorbents are cation exchange resins, anion exchange resins, Calcium Carbonate, clay, alumina and silica. The support media used include, ion exchange resins like DEAE – Sephadex and (CMC) Carboxymethyl Cellouse. The disadvantatges of this method includes the attachment of other particles along with enzymes, as it is a non-specific process. The other disadvantages would be the low loading of enzyme on a unit amount of surface along with the weak binding strength. However, the advantages that could be considered are, the easy procedure of immobilization, the reversible adsorption and no deactivation of enzymes due to adsorption.

5.5   Covalent Binding

In covalent binding the enzymes are adhered to the surfaces of the support by the formation of covalent bond. The functional group such as amino, carboxyl, hydroxyl and sulphydryl groups are responsible for binding of enzyme molecules to the support material. Also it is essential that these functional groups must not be present in the active site. The chemical agents used for activating the functional groups on support material are, gluteral dehyde carbodiimide and cyanogen bromide.

Some support materials with different functional groups and chemical reagents, used for covalent binding of proteins are given in table.

Table 5.1 Covalent binding of Enzymes to supports.

Supports with OH – OH

(a) Using cyanogen bromide

$$HC-OH \quad + CNBr \longrightarrow \quad HC-O{\Large\diagdown}_{\displaystyle C=NH} \quad + protein-NH_2 \longrightarrow \quad HC-O-CO-\underset{H}{N}-Protein$$

(b) Using S-triazine derivates

$$+ protein-NH_2$$

(a) By diazotization

Supports with – NH_2

$$-NH_2 \xrightarrow[HCL]{NaNO_2} \quad -N_2^+ \; Cl^- \xrightarrow{+ protein-NH_2} \quad -N=N-Protein$$

(b) Using glutaraldehyde

Supports with - COOH

(a) Via azide derivative

1.
$$-NH_2 + HCO-(CH_2)_3-HCO \longrightarrow -N=\underset{H}{C}-(CH_2)_3-HCO \xrightarrow{+ protein-NH_2} -N=\underset{H}{C}-(CH_2)_3-\underset{\underset{Protein}{|}}{\overset{H}{\underset{N}{C}}}$$

2.
$$-O-\underset{H_2}{C}-COOH \xrightarrow[H^+]{CH_2OH} -O-\underset{H_2}{C}-COOCH_3 \xrightarrow{+ protein-NH_2} -O-\underset{H_2}{C}-CO-\underset{H}{N}-NH_2$$

(b) Using carboiimide

$$-O-\underset{H_2}{C}-CO-\underset{H}{N}-NH_2 \xrightarrow[HCL]{NaNO_2} -O-\underset{H_2}{C}-CON_3 \xrightarrow{+ protein-NH_2} -O-\underset{H_2}{C}-CO-\underset{H}{N}-Protein$$

1.

$$-COOH + \underset{N-R}{\overset{N-R}{C}} \longrightarrow -\overset{O}{C}-O-\underset{N-R}{\overset{HN-R_1}{C}} \xrightarrow{+ protein-NH_2} -\overset{O}{C}-\underset{H}{N}-Protein + O=\underset{HNR}{\overset{HNR_1}{C}}$$

2.

Supports cotaining anhydrides

$$-CH_3-CH-CH-\underset{H_2}{C}- \quad + Protein-NH_2 \longrightarrow \quad HOOC-CH-\underset{H_2}{C}-$$
$$O=C \qquad C=O$$
$$\diagdown O \diagup$$
$$-CH_3-CH \qquad O=C-NH-Protein$$

The amino carboxyl or side groups (R) of the polypeptide chain are the binding groups on the protein molecule.

The another method of enzyme immobilization is the cross linking of enzyme molecule with each other by using by agents such as, 2, 2-disulfonic acid, bis-diazobenzidine and gluteral dehyde. This can be achieved by cross linking enzymes with gluteraldehyde forming insoluble aggregate and adsorbed enzyme may be cross linked. Cross linking may also achieved by impregnation of porous support material with enzyme solution. The disadvantage due to cross liking may be the significant changes in active site of enzymes and severe diffusional limitations. The enzyme and its particular application will be the deciding factor for choosing support material and immobilization method. Support material may be selected based on its binding capacity. Binding capacity inturn is function of charge density functional groups, porosity and hydrophobicity of the support surface. Support material may also be chosen based on its stability and retention of enzyme activity, which depends on functional groups on support material and micro environmental conditions. A loss in enzyme activity may be observed due to conformational changes on the enzyme or due to involvement reactive groups of active site in binding. Generally immobilization causes loss in enzymes activity and stability. Immobilization may also increase enzyme activity and enzyme stability in some cases when favourable micro-environmental conditions are existing.

5.6   Immobilization Enzymes Reactions

Inspite of the low enzyme activity, immobilized enzymes systems are useful for various process application. Mainly the advantages considered include, their amenability to use in continuous system, ease in removal of immobilized pellets for reuse or recycling and separation of enzyme free product in the final step.

The Immobilization enzyme systems are mostly used in continuous operations. The commonly used continuous type of bioreactors are (a) Plug flow reactors (b) Continuous stirred tank reactor (c) Packed bed reactor and (d) fluidized bed reactors.

In continuous stirred tank reactors fully mixed zones are available for proper interaction between the enzymes and substrate. The only disadvantage would be disruption of enzyme pellets due to shear force.

The bioreactors with low hydrodynamic shear will be useful. These bioreactors include, fluidized bed, air lift or packed column.

Fluidized bed bioreactors exhibit the good features of continuous stirred tank and packed bed bioreactor.

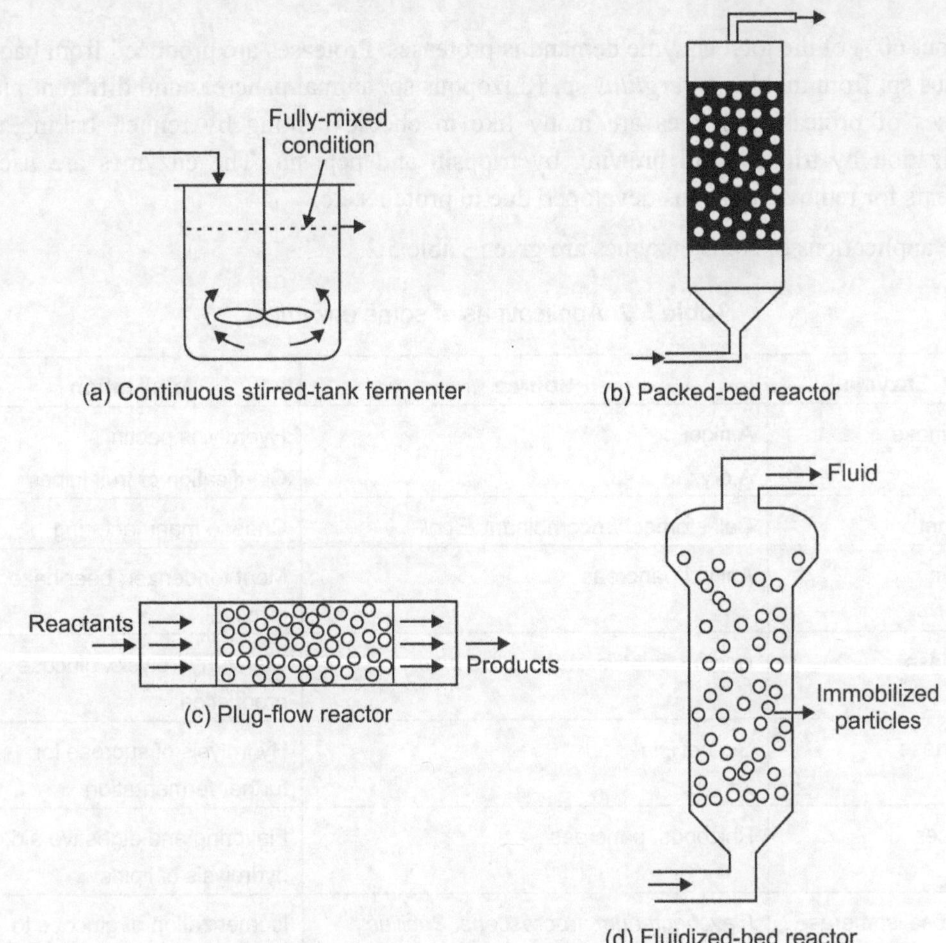

Fig. 5.2 Various types Immobilized reactors.

5.7 Applications of Enzymes

The advent of rDNA Technology (Recombinant DNA technology) opened many opportunities for producing various rare enzymes in large quantities at low cost. The demand for production of Chirally pure compounds in pharmaceutical sector is increasing. Chirality plays an important role in products for example, in a racemic mixture one enantiomer may have therauptic value than the other enantiomer which may cause side effect or may not possess therauptic value. The enzymes play a vital role in recognizing, chiral isomers and reacting with one of them for pharmaceutical products.

Microbial enzyme are finding various applications due to their ability to withstand different environmental conditions like high concentrations, pH, and temperature. A variety of enzymes will also be developed by the applications of recombinant DNA technology to different microorganisms.

About 60% of the total enzyme demand is proteases. Proteases are produced from bacteria Bacilllus sp. from molds, *Aspergillus* sp, Rhizopous sp. animal pancreas and different plants. The uses of protease enzymes are many like in cheese making by rennet, baking meat tenderization by trippsin and brewing by trippsin and peppsin. The enzymes are used in detergents for removal of stains developed due to proteins etc.

The applications of some enzymes are given Table 5.2.

Table 5.2 Applications of some enzymes.

Enzyme	Source	Application
Pectinase	A.niger A.oryzae	Hydrolysis pectin Clarification of fruit juices
Rennet	Calf stomach/recombinant *E.coli*	Cheese manufacturing
Trypin	Animal pancreas	Meat tenderizer, beer haze removal.
Amylase	*Bacillus subtilis, Aspergillus niger*	Starch hydrolysis, glucose production.
Invertase	*S.cerevisiae*	Hydrolysis of sucrose for further fermentation
Lipases	Rhizopus, pancreas	Flavoring and digestive aid, hydrolysis of lipids
Glucose isomerase	*Flavobacterium arborescens, Bacillus coagulans, Lactobacillus brevis.*	Isomerization of glucose to fructose

Pectinases are produced by *Aspergillus niger.* Pectinases contain, pectein esterage. Polygalacturonase and poly methyl galacturonase Pectinases are applied in improving yield of wine making, fruit juice processing and clearing the juice.

Lipases are produced from animal pancreas, molds and yeasts. Lipases convert lipds into fatty acids and glycerol. These enzymes are used in soaps manufacturing for hydrolyzing oils and in waste water streams for hydrolyzing the lipid-fat content. Lipases may be used in cheese and butter industry for imparting flavor. Lipases are also used in detergents. Amylases are produced from various microorganisms like *A.niger B.subtlis* and are used in hydrolysis of starch.

The three different amylases α-amylase, β-amylase and glucoamylase are frequently used in industries. α-amylase solubilizes amylose and is known as starch – liquefying enzyme. β-amylase known as saccharifying enzyme hydrolyses non-reducing ends of amylose to produce maltose. Glucoamylase hydrolyzes amylopectim portion of starch and also known as saccharifying enzyme. Glucose is produced by enzymatic hydrolysis of starch. Cellulase enzymes are produced from different microorganism like Trichoderma viride, Trichoderma reesci, *Aspergillus niger* and Thermomonospora. These cellulose enzymes hydrolyze cellulose to cellobiose. Cellulase are used in various industries like alcohol fermentation from biomass, brewing and waste water treatment process.

The enzyme pencillin acylase is used in the production of semisynthetic pencillins. It converts pencillin G to 6-amino pencillinic aicd (6 APA) and 6 APA is a precursor for semi synthetic penicillin derivatives.

Meat tenderization is carried out by the pectolytic enzymes namely papain. Invertase enzyme is used for conversion of sucrose to mixture of glucose and fructose in industries making confectionaries. Immobilized aspertase enzyme is used for conversion of fumarate to aspartate. Aspartate is further reacted with L-phenylalanine to produce aspartame, low caloric sweetener.

Enzymes are used for analytical purposes in food industries. In pasteurization of milk the degree of pasteurization is measure by absence of activity of phosphates and invertase enzyme. Also the food is tested for activity of microbial enzymes for possible contamination. The activity of microbial enzymes should not be present in sterile food immobilized enzymes are used in biosensors. Biosensors are used as diagnostic kits for identifying certain diseases like tuberculosis. Enzymes are used in detergents for removing stains. The protease enzymes produced for *Bacillus subtilis* are frequently used in detergents.

5.8   Control of Enzyme Activities

The control of enzyme activities plays an important role in understanding bioprocess. The enzyme activity could be controlled to over produce the microbial products of importance.

The control of pathways by cells is discussed for basic understanding, by considering a pathway of product P_1. The first reaction in the pathway is mostly inhibited by the liner accumulated product. The effectiveness of the enzyme is reduced due to binding of secondary site by the end product. The product P_1 in excess will deactivate the pathway and they substrate used to make P_1 can be utilized for other purpose.

This concept can be applied to the complicated pathways (Fig 5.3). For example P_1, P_2 could be essential metabolites where P_1, P_2 may be access to deactivate pathways, so for efficient utilizing substrates.

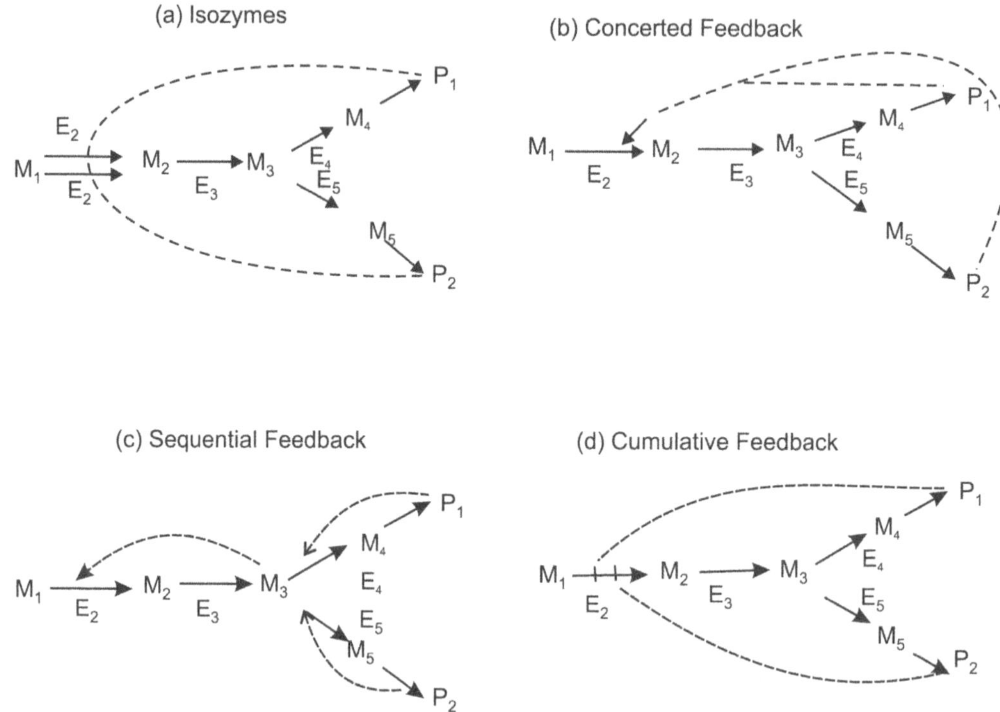

Fig. 5.3 Feed back control in branched path ways.

(a) Iso-functional enzymes (isoenzymes) may be used for inhibition of the pathways. Two separate enzymes will carryout the same conversion and different end product will be the inhibitor in each case. P_1 is added in excess quantity in growth medium , then it inhibits isoenzyme E_2 and E_2^1 is active. Eventually isoenzymes E_2 enhance synthesis of P_2.

(b) Concerted feed back inhibition: A single enzyme with allosteric binding sites (for P_1 and P_2) controls the pathway. The excess of P_1 or P_2 cannot inhibit E_2 enzyme but excess of P_1 and P_2 will fully inhibit E_2.

(c) Sequential feedback inhibition : It would be another approach. An intermediate at the branch point will be accumulated and inhibits the pathway. Excess of P_1 and P_2 inhibit E_4 and E_5. However, if E_4 and E_5 is blocked M_3 will accumulate. But if both E_4 and E_5 are blocked M_3 accumulation will be more. Intermediate flux levels will be allowed if P_1 or P_2 is high but pathway will be inactivated if both P_1 and P_2 are in excess.

(d) Cumulative feedback inhibition

A single allosteric enzyme may consist effective sites for several end products of a pathway and only partial inhibition is caused by each effective site of enzyme. All the effective sites will result in complete inhibition due to cumulative effect and the control is known as cumulative feedback inhibition or cooperative feedback inhibition will inhibition occurs at enzyme level and is raid, repression is the phenomena that occurs at genetic level. Repression is a slow process and it is also difficult to reverse the process.

It should be noted that the control strategy adopted by an organism for a particular pathway may differ from a closely related organism having similar pathway. The cellular regulatory strategy enables to choose optimal fermentor and operator strategy includes strain improvement.

5.9   Transport Molecules Across Cellular Membranes

To retain metabolic activity and to grow, the cell has to take nutrient from its extra cellular environment. The regulation of metabolic activity depends on which nutrients enter the cell and at what rate. The energy independent or energy dependent mechanism will be useful for molecules to enter the cell. Passive diffusion and facilitated diffusion are two examples of energy independent uptake. Active transport and group translocation are the examples of energy dependent uptake mechanism.

Passive diffusion

In passive diffusion molecules move from high to low concentration

$$J_A = K_P (C_{AE} - C_{AI}) \qquad \qquad .....(1)$$

Where J_A = flux of species 'A' across membrane

C_{AE} = Extracelluar concentration of species A (mol/ cm^3)

C_{AI} = Intracellular concentration

The cytoplasimic membrane consists of a lipid core. The permeability K_P is very low for charged or large molecules with very small flow of material across the membrane. Passive diffusion facilities the cellular uptake of water and oxygen. Lipids or highly hydrophobic compounds have high diffusion in cellular membranes due to passive diffusion.

Facilitated (Transport) diffusion

In facilitated diffusion, a carrier molecule (protein) combines specifically and reversibly with molecules. The membrane may be embedded with a carrier protein. Probably the carrier protein after binding the target molecule, undergoes conformational changes resulting in the release of the molecule on the intracellular side of the membrane. The carrier protein binds to the target molecules on the intercellular side of membrane and releases the molecule from the cell. So that net flux of a molecules depends on the gradient of concentration. The solubility target molecule is increased due to carrier protein. Since the binding of molecule to carrier protein can be saturated, the dependence of flux rate of the target molecule on the concentration will be different then as shown in equation (1).

The equation for the uptake of facilities transport is

$$J_A = J_{A\,max} \left\{ \left[\frac{C_{AE}}{K_{MT} + C_{AE}} \right] - \left[\frac{C_{A1}}{K_{MT} + C_{Az}} \right] \right\} \qquad .....(2)$$

Where

K_{MT} = The binding affinity of substrate of substrate (mol/cm^3)

J_{Amax} = Maximum flux rate of A (mol/cm^2-s)

The net flux will be into the cell when $C_{AE} > C_{Al}$

When

$C_{AE} > C_{Al}$, there will be a net flux A from the cell .

In eukaryotic cells, sugars and other low molecular weight organic compounds are diffused due to facilitated transport, but is not very common in prokaryotic cells. Nevertheless, the uptake of glycerol in E-coli is due to facilitated transport.

Active Transport

In active transport the proteins are embedded in cellular membrane like facilitate transport. The active transport across against a concentration gradients. The intercellular concentration of molecule is much more greater than the extra cellular concentration. The movement of molecule up concentration gradient is not favorable thermodynamically and will not happen spontaneously, so energy must be supplied. Various energy sources that can be applied in active transport are,

(a) Electrostatic or pH gradients of the proton motive force.

(b) Secondary gradients derived from the proton motive force by over active proton systems i.e., Na$^+$ or other ions and hydrolysis of ATP. The protonmotive force, results from the extrusion of hydrogen as protons. Such gradients are formed in the respiratory system of cells.

For active transport an equation analogus to Michaelis – Menten Kinetics can be written to describe uptake,

$$J_A = J_{A\,max}\left[\frac{C_{AE}}{K_{MT} + C_{AE}}\right] \qquad .....(3)$$

Equation 3 is apt, when the cell is in energy –sufficient state.

References

1. Boyer. P.D (Ed) 1970 – 1974, The Enzymes, 3rd Edu Vol 1 to 10 Academic Press, New York.

2. Nielson P.H et al (1994),"Enzymes applications (Industrial), Kick –other ECT, Kroschwitz, J.I (Executive Ed), 4th edu, Vol 9 John wiley, New York p 164.

3. Wang. D.l.C, Cooney C.L Demain, A.L, Dunnil, P., Humphrey, A.E. and Lilly, M.D (1979). Fermentation and enzyme technology, John wiely and sons New York.

Review Questions

1. Describe enzyme classification.
2. Write about enzyme specificity.
3. Describe methods of Immobilization.
4. What is immobilization?
5. Write various applications of enzymes.
6. Describe about, passive, facilitated and active transport.
7. Describe control of enzyme activities.

CHAPTER 6

Enzymes

Enzymes are basically proteins. The molecular weight of proteins vary from 15000 dalton to several million daltons. Enzymes are specific in nature and the reaction rates of enzyme catalyzed reaction are much higher than the chemically catalyzed reactions. Hitherto more than 2000 enzymes are known. Enzymes are generally named either by adding ase to the substrate or by adding –ase to the reaction catalyzed. Ex: urea – urease and alcohol dehydrogenation – alcohol dehydrogenase. Hydrolytic enzymes may have simple structure like folded polypeptide chain. Enzymes contain non-protein group like either cofactor such as metal ions, Mg, Zn, Mn, Fe, or coenzymes like NAD, FAD, CoA, vitamins or complex organic molecule. The enzyme that contain non-protein group is called holoenzyme, while the protein group is called apo-enzyme. So holoenzyme = apo-enzyme + cofactor. The major classification of the enzyme is shown in following Table 6.1.

Table 6.1 The major classification of the enzyme.

1. Oxido-reduction : Oxidation reduction Reaction

    Acting on – CH – OH

    Acting on – C = O

    Acting on – C = CH

    Acting on – CH – NH$_2$

    Acting on – CH – NH

    Acting on – NADH; NADPH.

2. Transferase – Transfer of functional groups

    One – carbon group

    Aldehydic or Ketonic groups

    Acyl groups

    Glycosyl group

    ....

    2.7 Phosphate groups

    2.8   S-containing groups

Table 6.1 *Contd....*

3. Hydrolases – hydrolysis reactions

    Esters

    Glycosidic bonds

    ....

    3.4 Peptide bonds

    3.5 C- N bonds

    3.6 Acid anhydride

4. Lyases – addition to double bonds

    C = C

    C = O

    C = N

    ....

5.0 Isomerase – Isomerisation

5.1 Racemases

    ....

6. Ligases – formation of bonds with ATP cleavages

    6.1 C – O

    6.2 C-S

    6.3 C-N

    6.4 C-C

6.1 Function of Enzymes

The activation energy of the reaction catalyzed in lowered by the enzymes by binding to the substrate and then forming enzyme substrate complex. Free energy change or the equilibrium constant is not affected by the enzyme but activation energy will be affected by the enzyme but activation energy will be altered. Action of the enzyme in relation to the activation energy is shown in Fig. 6.1.

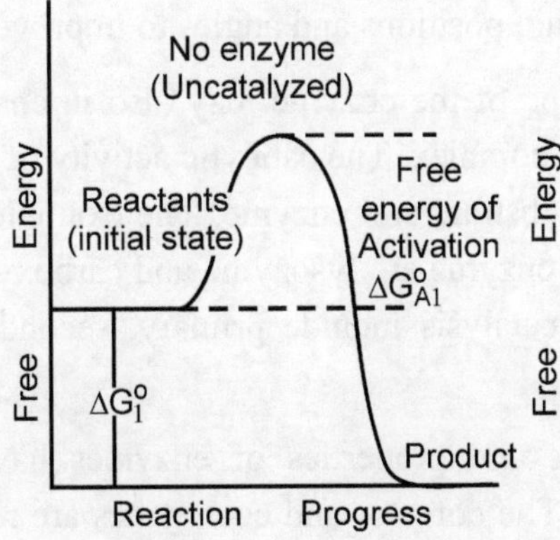

Fig. 6.1 Activation energy of enzymatically uncatalyzed reactions.

The decomposition of hydrogen peroxide and the activation energy required are well studied for different types of required and well studied for different types of catalysis. Now, the activation energy for uncatalyzed reaction will be around 18 Kcal/ mol at 20^0C where as the activation energy would be 13 Kcal/mol at 20^0C for chemically catalyzed reactions and 7kcal/mol at 20^0C for enzymatic ally catalyzed reactions. (by catalase).

The formation of enzyme substrate complex directs the lowering of activation energy. The enzyme and substrate complex is formed due to weak forces like Vanderwaals forces and hydrogen binding. The active site of the enzyme is used for the binding of the substrate on it. The enzyme molecule is bigger than the substrate molecule. And the enzyme is like a lock whereas the substrate is like a key. So the binding can be represented by lock and key model shown in the Fig. 6.2.

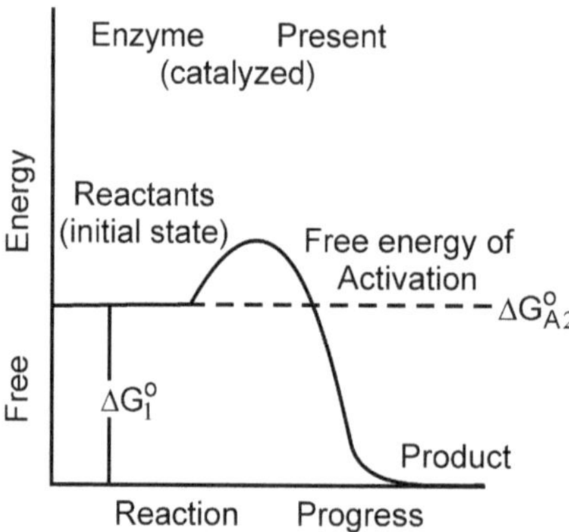

Fig. 6.2 Activation energies of enzymatically catalyzed reactions.

In commonly known as proximity effect enzymes hold, multiple substrate whose reactive regions close to each other at their active site. In the concept known as orientation effect, enzymes hold substrate at certain positions and angles to improve the reaction rate.

The three dimensional shape of the enzyme may also in change. Some cases during the formation of substrate enzyme complex. The catalytic activity of enzyme may be affected due to this sort of induced fit of substrate and enzyme. The examples for this type of change of three dimensional structure of enzyme are lysozyme and carboxy peptidase-A. The structures responsible for the enzymes catalysis include primary, secondary, tertiary and quaternary structure.

The folding characteristics and properties of enzymes have significant effect on the catalytic activity of enzymes. The cofactor and coenzymes are required by some enzyme for their proper functioning. A list of cofactors and coenzymes of some enzymes are in Table 6.2.

Table 6.2 Cofactors and Coenzymes of some enzymes.

	Coenzyme	*Entity transferred*
Zn^{2+}	Nicotinamide adenine dinucloeotide	Hydrogen atoms (electron
Alcohol dehydrogenase	Nicotinamide adenine dinucloeotide	Hydrogen atoms (electron
Carbonic anhydrase	Phosphate	Hydrogen atoms (electron
Carboxy peptidase	Flavin mononucleotide	Hydrogen atoms (electron
Mg^{2+}	Flavin adenine dinucleotide	Hydrogen atoms (electron
Phosphohydrolases	Coenzyme Q	Aldehydes
Phosphotransferases	Thiamin pyrophosphate	Acyl groups
Mn^{2+}	Coenzyme A	Acyl groups
Arginase	Lipoamide	Alkyl groups
Phosphotransferases	Cobamide coenzymes	Carbon dioxide
Fe^{2+} or Fe^{3+}	Biocycin	Amino groups
Cytochromes	Pyridoxal phosphate	Methyl, methylene, form
Peroxidase	Tetrahydrofolate coenzymes	or formimino groups
Catalase		
Ferrodoxin		
Cu^{2+} (Cu^{+})		
Tyrosinase		
Cytochrome oxidase		
K^{+}		
Pyrovate kinase		
(also requires Mg^{2+})		
Na^{+}		
Plasma membrane ATPase (also requires K^{+} and Mg^{2+})		

6.2 Kinetics of Enzyme Catalyzed Reaction

VCR. Henri (1902) and L.Michaelis and M.L Menten (1913) developed a mathematical model for single substrate enzyme catalyzed reaction. Simple enzyme catalyzed Kinetic reactions are known as Michaelis – Menten Kinetics or saturation Kinetics.

The effect of enzyme catalyzed reaction on substrate concentration is given in Fig. 6.3. The Fig.6.3 is plotted with the data obtained from the batch reactor with constant volume where initial substrate concentration $[S_0]$ and enzyme concentration $[E_0]$ are known. In biological systems multi-substrate – multi-enzyme reactions do normally take place. Also the fixed number of active sites are available in enzyme solution where the substrate can find. All the active sites will be occupied with substrate or the sites would be saturated kinetics could be obtained from the saturation kinetics, could be obtained from the formation of the enzyme substrate complex reversibly and also further dissociation of enzyme substrate complex product plus enzyme.

$$E + S \xrightarrow[k_1]{k_1} ES \xrightarrow{k_2} P + E \qquad \ldots\ldots(1)$$

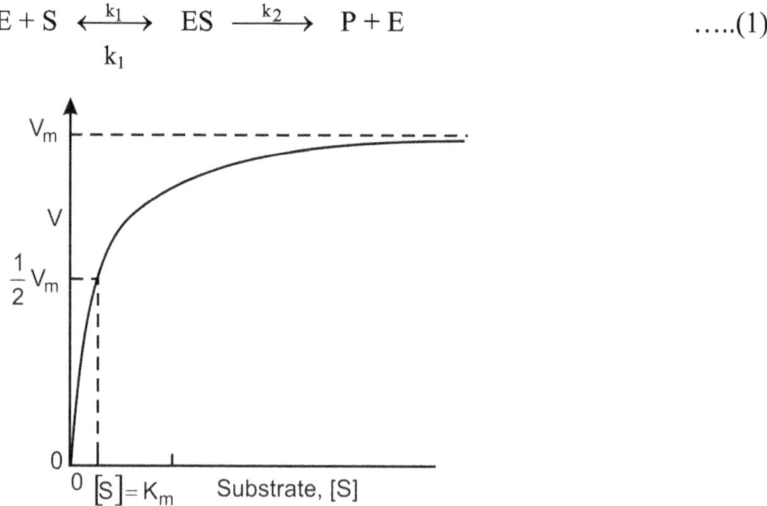

Fig. 6.3 Effect of substrate concentration on the rate of enzyme catalyzed reaction.

In the above reaction it is assumed that the rate of Enzyme substrate complex is rapid and also the reversible reaction of the second step is negligible. The assumption is irreversible second step holds good only when the product accumulation is negligible in the beginning of the reaction.

For developing a rate expression for the enzyme catalyzed reactions two approaches are followed

1. Rapid equilibrium approach
2. Quasi steady state approach

Simple Enzyme Kinetics

The rate of expression development for the mechanism of equation (1), have the same initial steps.

So the rate of product formation is

$$\upsilon = \frac{d[P]}{dt} = K_2[ES] \qquad \ldots\ldots(2)$$

υ = rate of product formation or substrate consumption i.e moles/l-s

The constant K_2 can be denoted as k_{cat} per biological systems.

The rate of reaction of ES complex is given as

$$\frac{d[ES]}{dt} = K_1[E][S] - K_{-1}[ES] - K_2[ES] \qquad \ldots\ldots(3)$$

As the enzyme is not consumed, the conservation equation on the enzyme gives as,

$$[E] = [E_0] - [ES] \qquad \ldots\ldots(4)$$

Now to achieve an analytical solution an assumption will be made.

Rapid Eqilibrium Assumption

This approach was used by Henri and Michaelis and Menten.

The equilibrium coefficient can be used to express [ES] in terms of [S] by assuming a rapid equilibrium enzyme and substrate to form [E] complex.

So the equilibrium constant is

$$K_m^1 = \frac{K_{-1}}{K_1} = \frac{[E][S]}{[ES]} \qquad \ldots\ldots(5)$$

Now if enzyme is conversed

$$[E] = [E_0] - [ES],$$

So

$$[ES] = \frac{[E_0][S]}{(K_{-1}/K_1) + [S]} \qquad \ldots\ldots(6)$$

$$[ES] = \frac{[E_0][S]}{K_m^1 + [S]} \qquad \ldots\ldots(7)$$

Where

$K^1_m = K_{-1}/K_1$ and this is the dissociation constant of ES complex.

So substitution of equation (7) in (2) we have,

$$\upsilon = \frac{d[p]}{dt} = K_2 \frac{[E_0][S]}{K_m^1 + [S]} = \frac{V_m[S]}{K_m^1 + [S]} \qquad \ldots\ldots(8)$$

and $V_m = K_2[E_0]$.

Here the maximum forward velocity of the reaction is V_m .

While the addition of more substrate has no influence on V_m. V_m do change when more enzyme is added .K_m^{1} is known as Michaelis – Menten equation and prime indicates that it is derived by assuming rapid equilibrium in the first step. Enzyme high affinity for substrate is

Known by the low value of K^1_m, K_m^1 relates to the substrate concentration , giving half of the maximum reaction velocity.

Similar equation (8) can be derived by assuming more general approach i.e., the quasi – steady state assumption.

6.3   The Quasi – Steady State Assumption

Even through the enzyme substrate reaction shows saturation type kinetics, rapid equilibrium assumption may not hold good. The quasi steady state assumption was first proposed by GIE. Biggs and J.B.S Haldane. It is observed that in most of the systems like closed say batch reaction is used where initial substrate concentration exceeds initial enzyme concentration.

However $[E_0]$ is small $d[ES]/dt = 0$ does not hold good. In actual time come as represented by equation time as represented by equation (2) to (4), it is shown that in a closed system the quasi – steady state holds after a brief transient as shown in Fig (6.4) at $[S_0] \gg [E_0]$

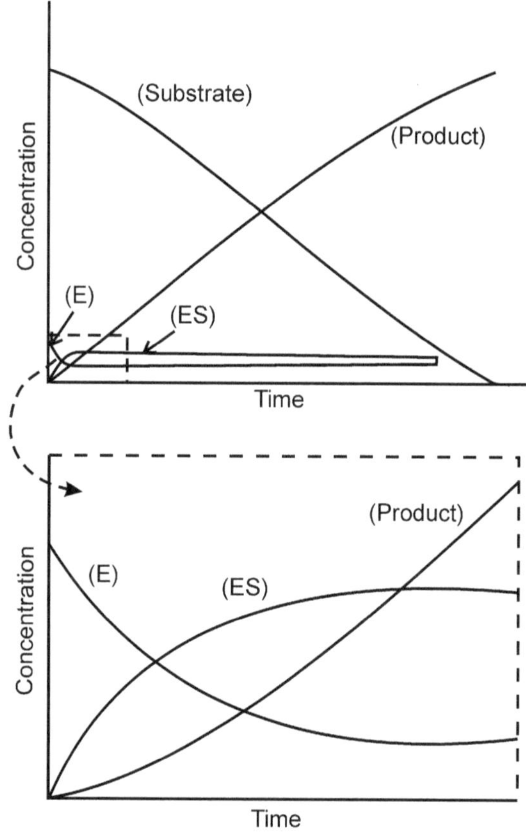

Fig. 6.4 Time course of the formation of an enzyme/substrate complex and intiation of steady state. ➤ magnified path.

Now by applying the quasi steady state assumption to equation (3) we have

$$[ES] = \frac{K_1[E_0][S]}{K_{-1} + K_2} \qquad \qquad .....(9)$$

Substituting enzyme conservation eqn (4) in eqn (9) gives

$$[ES] = \frac{K_1\{[E_0] - [ES][S]\}}{K_{-1} + K_2} \qquad \qquad .....(10)$$

Upon solving eqn (10) for [ES] we have

$$[ES] = \frac{[E_0][S]}{\dfrac{K_{-1} + K_2}{K_1} + [S]} \qquad \qquad .....(11)$$

Now substituting eqn (11) into eqn (2), we get

$$\upsilon = \frac{d[P]}{dt} = \frac{[E_0][S]}{\dfrac{K_{-1} + K_2}{K_1} + [S]} \qquad \qquad .....(12)$$

$$\upsilon = \frac{V_m[S]}{K_m + [S]}$$

Where

$$K_m = \frac{K_{-1} + K_2}{K_1}$$

$$V_m = K_2[E_0]$$

Normally it would be difficult to know whether K_m or K_m^1 is more appropriate.

6.4   Estimation of Michaelis – Menten Parameters

K_m and V_m can be determined by conduction initial rate experiments of course high precision may not be possible.

The experiments conducted could be like this known amount of substrate of $[S_0]$ and enzyme $[E_0]$ is fed to the batch reactor. Substrate or product concentration will be plotted against time. The initial slope of the curve could be estimated

$\vartheta = d[P]/dt$ at t = 0, υ depends on E_0 and S_0 υ and [S] data may be generated by conducting various experiments could be plotted in. (Fig 6.3). How ever K_m may be not be determined accurately. So other methods like Line Weaver Burk plot may be used.

Line Weaver Burk Plot or Double Reciprocal Plot

Equation (13) may be written as,

$$\frac{1}{\vartheta} = \frac{1}{V_m} + \frac{K_m}{V_m}\frac{1}{[S]} \qquad\qquad .....(13)$$

Now $1/\upsilon$ versus $1/[S]$ is plotted which gives a linear line with a slope of K_m/V_m.

The Y- axis intercept would be $1/V_m$

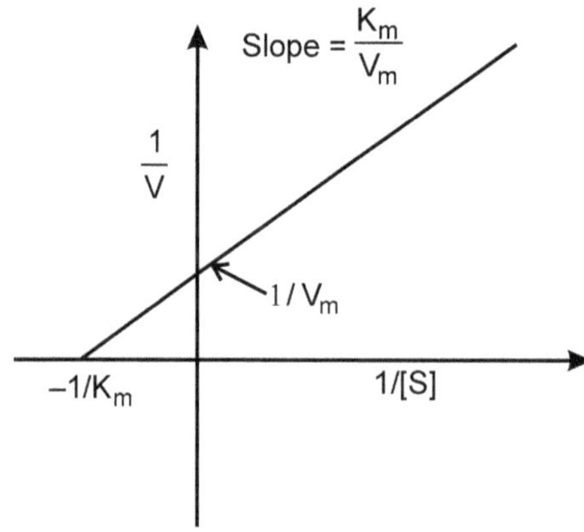

Fig. 6.5 Line Weaver – Burk Plot.

With the plot V_m may be estimated accurately but may not be estimated accurately due to error about reciprocal of a data point which is not symmetric.

Also the data points at low substrate concentration may influence slope and intercept more than those at high substrate concentrations.

Eadie – Hofstee Plot

Equation (12) may be written as

$$\vartheta = V_m - K_m \frac{\vartheta}{[S]} \qquad\qquad .....(14)$$

So υ versus V/[S] plot gives a line with a slope $-K_m$ and intercept Y axis at V_m . There plots may have large errors due to presence of υ both coordinates.

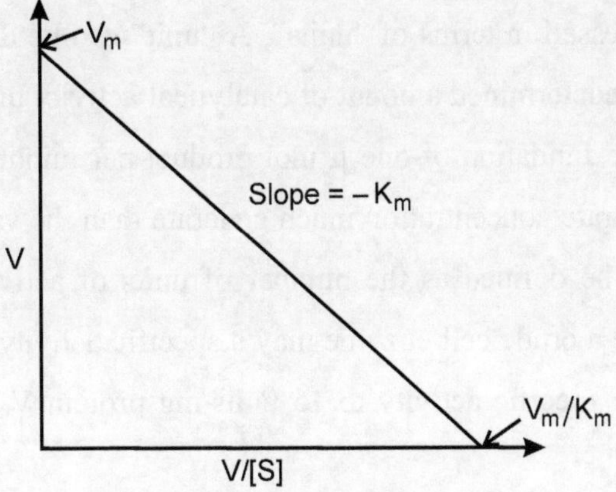

Fig. 6.6 Eadie – Hofstee plot.

Hanes – Woolf Plot

Equation (12) may be written as

$$\frac{[S]}{\vartheta} = \frac{K_m}{V_m} + \frac{1}{V_m}[S] \qquad\qquad .....(15)$$

[S]/V versus [S] plat gives a line with a slope $1/V_m$ and intercepts Y-axis at K_m/V_m wide

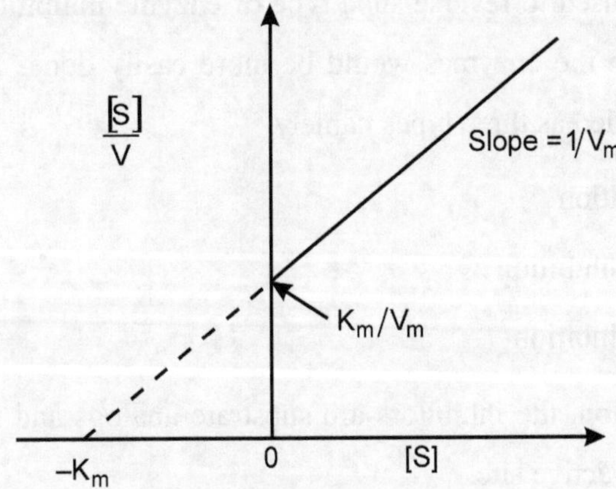

Fig. 6.7 Hanes – Woolf Plot.

V_m may be determined accurately by using the plot.

Parameters of V_m of K_m

K_m is a function of rate parameters only and changes with pH or temperature. While V_m is a function of K_2 the rate parameters and initial enzyme concentration $[E_0]$. $[E_0]$ can be expressed in terms of mol/l or g/l for highly purified enzyme preparations. For crude enzymes the concentration is expressed in terms of "units". A "unit" can be defined as the amount of enzyme which gives a predetermined amount of catalytical activity under specific conditions. Also one unit may be the fundation of one μ mol product per minute at a specified pH and temperature when a substrate concentration much generate than the value of K_m. The specific activity of enzyme may be defined as the number of units of activity per amount of total protein. Now for instance a crude cell enzyme may a specific activity of 0.3 units/mg protein and after purification the specific activity to 15 units/mg protein V_m is expressed as μ mol product/ml-min.

6.5   Enzyme Inhibition

Enzyme inhibitors are is those compounds which bind to the enzymes and reduce the activity of the enzymes. The enzymes inhibitors may be reversible or irreversible. A stable complex is formed in between enzyme and irreversible inhibitors like heavy metals namely lead mercury, cadmium and others. However chelating agents like (EDTA) ethylene diamine tetra acetic acid and citrate, may be used to reverse such type of enzyme inhibition. The dissociation of reversible inhibitors from the enzymes would be more easily done. The reversible enzyme inhibitors could be classified as three types namely

1.   Competitive inhibition

2.   Non-competitive inhibition

3.   Un-competitive inhibition

In competitive inhibition, the inhibitors are substrate analogs and they compete with the substrate for the enzymes active site.

The competitive inhibition can be described as,

$$E + S \underset{k}{\longleftrightarrow} ES \xrightarrow{K_2} E + P$$

$$+$$

$$I \qquad\qquad\qquad\qquad \ldots..(16)$$

$$\downarrow k$$

$$EI$$

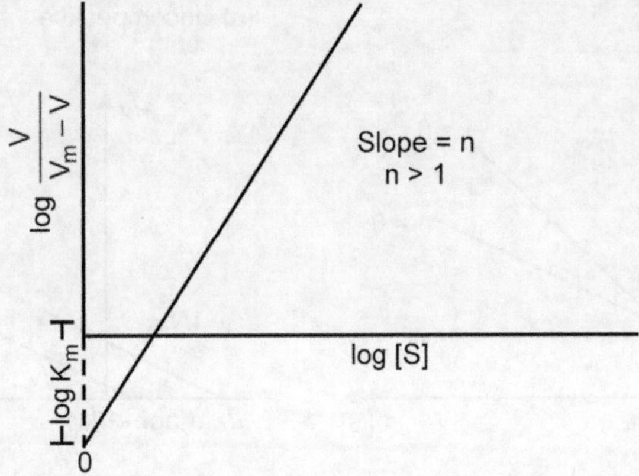

Fig. 6.9 Determination of cooperativity coefficient.

Now rapid equilibrium is assumed

$$K_m^1 = \frac{[E][S]}{[ES]}, K_I = \frac{[E][I]}{[E][I]} \qquad .....(17)$$

$$[E_0] = [E] + [ES] + [ES]$$

$$V = K_2\,[\,ES]$$

Equation for the rate of enzymatic conversion can be developed

$$\vartheta = \frac{V_m[S]}{K_m^1\left[1 + \dfrac{[I]}{K_I} + [S]\right]} \qquad .....(18)$$

Or

$$\vartheta = \frac{V_m[S]}{K_{m\,app}^1 + [S]} \qquad .....(19)$$

where $K_m^1 = K_m\,[1 + [I]/K_I]s$

In competitive inhibition $K_{m\,app}^1$ is increased and the reaction rate is reduced. If the substrate concentration is increased, the competitive inhibitor could be avoided. Competitive inhibition is shown in Fig. 6.10.

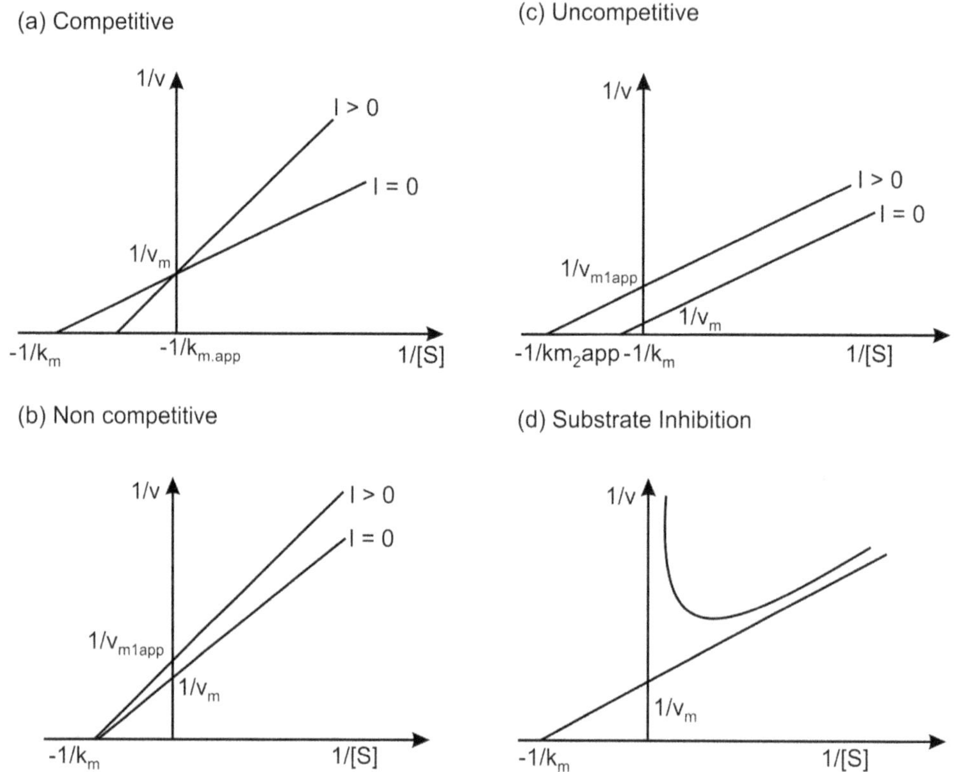

Fig. 6.10 Different forms of inhibited enzyme kinetics.

Non-competitive Inhibition

Non-competitive inhibitors bind enzyme sites which are not active site resulting in reduction of enzyme affinity to substrate. Also the non-competitive inhibitors are not substrate analogs.

Non-competitive inhibition can be described as,

$$E + S \leftrightarrow ES \rightarrow E + P$$

$$+ \qquad k_m^1 \qquad k_2$$

$$I$$

$$\updownarrow$$

$$EI + S \leftrightarrow ESI \qquad\qquad .....(20)$$

$$k_m^1$$

Now

$$K^1_m = \frac{[E][S]}{[ES]} = \frac{[EI][S]}{[ESI]}, K_I = \frac{[E[[I]}{[EI]} = \frac{[ES][I]}{[ESI]}$$

$$[E_0] = [E] + [ES] + [EI] \text{ s}$$

And

$$\upsilon = K_2 = [ES]$$

The following rate equation can be developed

$$\vartheta = \frac{V_m}{[1+\dfrac{[I]}{K_1}][1+\dfrac{K_m^1}{[S]}]} \qquad\qquad .....(21)$$

Or

$$\vartheta = \frac{V_m}{[1+\dfrac{K_m^1}{[S]}]} \qquad\qquad .....(22)$$

$$V_{m,app} = V_m / [1+[I]/K_1]$$

V_m is reduced as a result of non-competitive inhibition. Non-competitive inhibition could not be avoided due to high substrate concentration V_m is reduced and K_m^1 is increased in some form of non-competitive inhibition. This is expected when ESI can form a product.

Un-competitive Inhibition

In un-competitive inhibition the inhibitors do bind to ES complex rather than enzyme.

Un-competitive inhibition can be written as given below

$$E + S \overset{k_m^1}{=} ES \xrightarrow{\ k_2\ } E + P$$
$$+$$
$$I\ \ k_1$$
$$ESI \qquad\qquad .....(23)$$

By definition

$$K_m^1 = \frac{[E][S]}{[E]}, K_1 = \frac{[ES][I]}{[ESI]} \qquad\qquad .....(24)$$

$$[E_0] = [E] + [ES] + [\,ESI]$$

$$\upsilon = K_2\ ES$$

The rate of equation may be developed and may be given by

$$\vartheta = \frac{\dfrac{V_m}{1+\dfrac{[I]}{K_1}}[S]}{\dfrac{K_m^1}{1+\dfrac{[I]}{K_1}}[S]} \qquad\qquad .....(25)$$

Or

$$\vartheta = \frac{V_{mapp}[S]}{K^1_{mapp} + [S]} \qquad \qquad .....(26)$$

V_m, K_m^1 values will be reduced during

Uncompetitive Inhibition

V_m reduction has more prominent effect, then the K_m^1 reduction and this results in reduction of reaction rate. Un-competitive inhibition is shown in Fig. 6.10.

Substrate Inhibition

In substrate inhibition the enzyme reactions will be inhibited by high substrate concentrations.

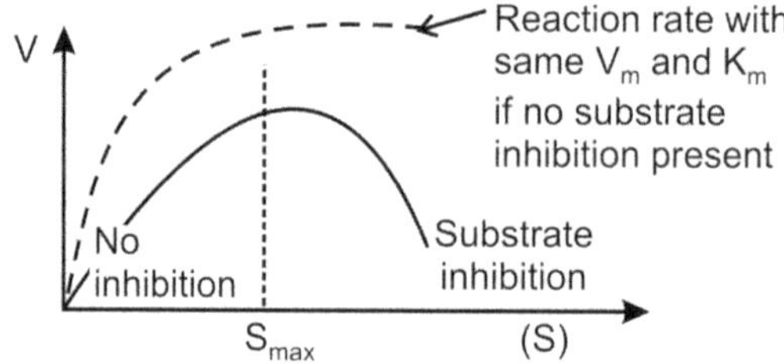

Fig. 6.11 Comparison of substrate inhibited and un-inhibited enzymatic reactions.

Uncompetitive substrate inhibition is given by

$$E + S \leftrightarrow ES \xrightarrow{\;K_2\;} E + P$$

$$+$$

$$S$$

$$I \ k_{s1}$$

$$ES_2$$

By definition

$$K_{s1} = \frac{[S][ES]}{[ES_2]}, K_m^1 = \frac{[S][E]}{[ES]} \qquad .....(27)$$

Rapid equilibrium assumption gives

$$\vartheta = \frac{V_m[S]}{K_m^1 + [S] + \dfrac{[S]^2}{K_{s1}}} \qquad .....(28)$$

Substrate inhibition is shown in Fig. 6.10

The inhibition effect is not observed at low substrate concentrations

i.e $[s]^2/K_s \ll 1$

The rate would be

$$\vartheta = \frac{V_m}{1 + \dfrac{K_M^{\ 1}}{[S]}} \qquad\qquad .....(29)$$

Or

$$\frac{1}{\vartheta} = \frac{1}{V_m} + \frac{K_m^1}{V_m}\frac{1}{[S]} \qquad\qquad .....(30)$$

A plot of $1/\upsilon$ Versus $1/[s]$ gives a line with a slope of K^1_m/V_m and $1/V_m$ intercepts Y axis
So

$$\upsilon = \vartheta = \frac{V_m}{1 + \dfrac{[S]}{K_{s1}}} \qquad\qquad .....(31)$$

$$\frac{1}{\vartheta} = \frac{1}{V_m} + \frac{[S]}{K_s , V_m} \qquad\qquad .....(32)$$

Now a plot of $1/\upsilon$ versus $[S]$ gives a line with a slope of $1/K_{s1}$, V_m and intercept $1/V_m$

6.6   pH and Temperature Effect on Enzymes

Active sites of certain enzymes contain ionic groups in a suitable form like acid or base. The ionic form of the active site of enzyme changes due to changes in pH of the medium, which results in changes in enzyme activity and also changes in reaction rate. The three dimensional shape of the enzyme may also be changed due to change in the pH. So it is known that the enzymes are active only at certain pH ranges. The maximum reaction rate K_m and stability of the enzyme may be affected due to the pH of the medium. Also the substrate may contain ionic groups and the pH of the medium may effect the affinity of substrate to the enzyme. pH dependence of the enzymatic reaction rate for ionizing enzymes are given by,

$$
\begin{array}{l}
E^- + H^+ \\[2pt]
\quad\uparrow\!\downarrow \qquad\quad k_m^{\ 1} \qquad\quad k_2 \\[2pt]
EH + S \; = EHS \longrightarrow EH + P \\[2pt]
\quad + \\[2pt]
\quad H^+ \\[2pt]
\quad\uparrow\!\downarrow \\[2pt]
\quad EH_2^{\ +} \qquad\qquad\qquad\qquad\qquad .....(33)
\end{array}
$$

$$K_m^1 = \frac{[EH][S]}{[EHS]}$$

$$K_1 = \frac{[EH][H^+]}{[EH_2^+]} \qquad \qquad .....(34)$$

$$K_2 = \frac{[E^-][H^+]}{[EH]}$$

$$[E_0] = [E^-] + [EH] + [EH_2^+] + [EHS]$$

$$\upsilon = K_2\,[EHS]$$

The rate expression may be divided

$$\vartheta = \frac{V_m[S]}{K_m^1\left(1 + \dfrac{K_2}{H^+} + \dfrac{[H^+]}{K_1}\right) + [S]} \qquad \qquad .....(35)$$

Or

$$\vartheta = \frac{V_m[S]}{K_{mapp}^1 + [S]} \qquad \qquad .....(36)$$

So the pH optimum of the enzyme in between Pk_1 and Pk_2.

The rate expression may be developed for the ionizing substrate

$$SH^+ + E \; \xrightarrow{\; k_m' \;} \; ESH \; \xrightarrow{\; K_2 \;} \; E + HP^+$$
$$\updownarrow k_1$$
$$S + H^+ \qquad \qquad .....(37)$$

$$\vartheta = \frac{V_m[S]}{K_m^1\left(1 + \dfrac{K_2}{[H^+]}\right) + [S]} \qquad \qquad .....(38)$$

Fig (6.12) shows the variation of enzyme activity with pH. The pH optimum for an enzyme is normally determined by conducting experiments.

Effect of Temperature

With the increase of the temperature rate of enzyme, catalyzed reactions also will increase up to a certain level. The enzyme denaturation takes place above a certain increased temperature, and enzyme activity decreases.

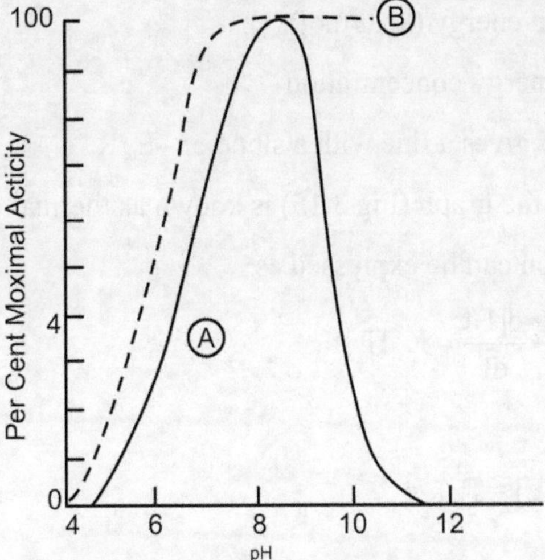

Fig. 6.12 The pH activity profiles of two enzymes (A) approximate activity for trypsin (B) approximate activity for cholinesterase.

Fig 6.13 shows the temperature effect on reaction rate. The descending portion of the curve is due to thermal denaturation. The nature of plot will depend on the length of time the reaction mixture is exposed to the test temperature.

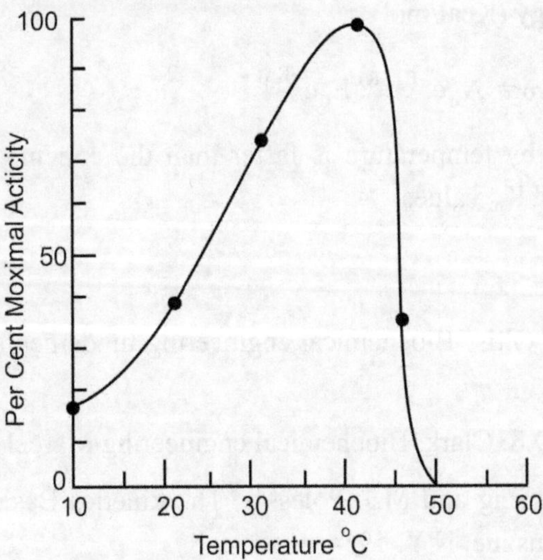

Fig. 6.13 Effect of temperature on activity of an enzyme.

The reaction rate varies as per Arrhenius equation during ascending part i.e temperature activation

$$\upsilon = K_2 [E] \qquad \qquad \dots(39)$$

$$K_2 = A\, e^{-Ea/RT}$$

Where E_a = activation energy (kcal/mol)

[E] = active energy concentration

A graph of υ versus $1/T$ gives a line with a slope of $-E_a/R$.

The descending part of the graph (Fig 3.15) is known as thermal denaturation

The thermal denaturation can be expressed as

$$-\frac{d[E]}{dt} = K_d[E] \qquad\qquad .....(40)$$

Or

$$[E] = [E]\ e^{-k_d t} \qquad\qquad .....(41)$$

Where

E_0 = initial concentration

K_d = denaturation concentration

K_d change with temperature as per Arrhenius equation

$$K_d = A_d e^{-E_a/RT} \qquad\qquad .....(42)$$

E_d = deactivation energy (Kcal/mol)

$$\upsilon = A_d e^{-E_a/RT} E_0 e^{-K_d t} \qquad\qquad .....(43)$$

Enzyme denaturation by temperature is faster than the enzyme activation. Variation of temperature effect V_m and K_m values.

References

1. Bialy, J.E and D.F Ollis 'Biochemical engineering fundamentals, 2nd ed, McGraw Hill Book, N.Y, 1986.

2. Blanch, H.W and D.S. Clark, Biochemical engineering Marcel Dekker, lnc, N.Y. 1996

3. F.H. Johnson, H.Eyring and M.J. Polessar, The kinetics Basis of Molecular Biology, John Wiely and Sons inc, N.Y. 1954.

4. K.J. Ladler and P.S. Bunting "The chemical Kinetics of enzyme Action, 2nd ed Oxford University Press London, 1973.

5. Moran, L.A, K.G. Scrimgeour, H.R. Horton, R.S. Ochs and J.D. Rawn, Biochemistry, Prentice – Hall, lnc, upper saddle River, N.J, 1994

Review Questions

1. Writer about enzyme classification.

2. Explain function of Enzymes.

3. Describe kinetics of enzyme catalyzed reaction.

4. Write about simple enzyme kinetics.

5. Explain rapid equimolar assumption.

6. Write about quasi steady state assumption.

7. How Michaelis – Menten parameters are evaluated? Explain.

8. Write about enzyme inhibition.

9. Explain competitive, non-competitive, un-competitive and substrate inhibition.

CHAPTER 7

Bioreactor Design

7.1 Design and Analysis of Biological Reactor

Bioreactor design is essential for developing a product of commercial significance. The laboratory reactors are operated at optimum conditions like mixing and flow characteristics. The reactor size and scale will be considered here and its effect on flow, mixing, mass and heat transfer patterns with the bioreactor.

The bioreactor systems need special attention due to the involvement of microorganisms as bio-catalysts and free enzymes of microbial origin.

The kinetics of the bioreactions depend on the cell growth of microorganisms and rate of substrate consumption.

The morphology of the cells also gives the overall reaction. Bioenergetics, stoichiometry, batch and continuous sterilization also play an important role during the design of bioreactors.

7.2 Different Types of Bioreactors

The different types of bioreactors are used in biotechnology process include, batch bioreactors, continuous stirred tank reactors, fed batch bioreactor, air lift bioreactor and fluidized bed bioreactor. The general description of each type is given here.

Batch Bioreactor

All the reactants are fed to the batch bioreactor initially during startup of the bioreactors. During the reaction there will not be any in-flow or out-flow from the system. These batch bioreactors are frequently used for small scale operations, for producing expensive chemicals and for developing new products. However, most of the pharmaceutical products are made by using batch type bioreactors only.

Continuous Reactors

Continuous reactors normally mean continuous flow stirred tank reactors (7.1). The mixing will be good due to vigorous stirring and uniform temperature could be maintained. The inflow and out-flow will be continuous and, product can be withdrawn continuously. Product quality could be controlled by online analysis of the out-flow stream, Mostly for continuous cultivation of organisms, these types of bioreactors are used.

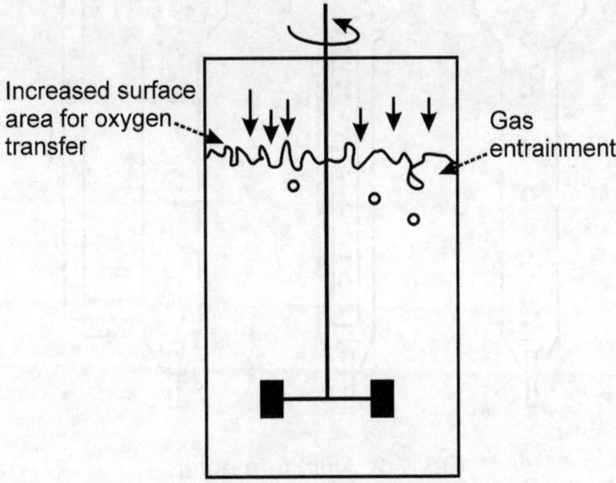

Fig. 7.1 Continuous stirred tank reactor.

Fed Batch Bioreactors

These types of bioreactors are used frequently for improving yields of various products. The nutrients, substrates are depleted during biochemical reactions. So, if the nutrients and substrate are fed continuously to the bioreactor, probably the product yield may be enhanced. The product will be removed after the batch is over. One of the advantages of these types of reactors is that the concentration of the limiting substrate may be maintained at a very low level, thus avoiding the repressive effects of high substrate concentration. The production of bakers yeast was achieved by using fed batch bioreactors.

Air Lift Bioreactors

In these reactors (7.2), mixing and internal circulation is achieved by passing air through the reactor contents. Air is passed through the spargers at high velocities into the riser. The difference of density between the liquid column in the riser and the liquid column in the down comer, will be the driving force for circulation of the medium in the air lift bioreactor, Circulation times in loops of 45 m height may be 120 seconds. These types of bioreactors may be used for producing SCP (single cell protein).

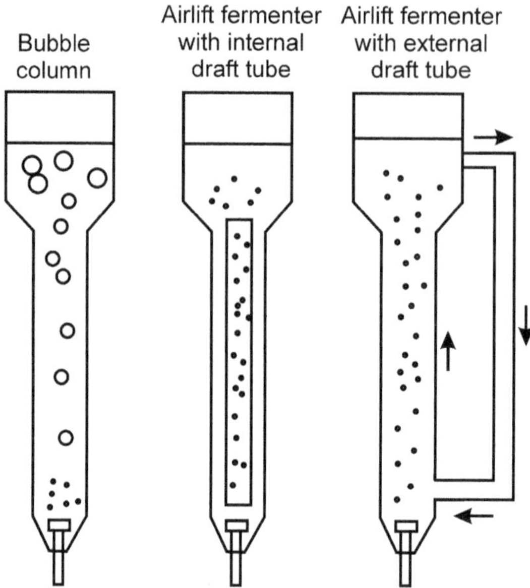

Fig. 7.2 Airlift fermenter.

Bubble Column Bioreactors

The bioreactors with large aspect ratio that is height to diameter ratio may be known as bubble column bioreactors. Mixing in these types of bioreactors is achieved by forcing compressed gas through the reactor contents. The operating costs of these reactors are low due to low energy requirement. These bioreactors may be operated either in batch mode or in continuous mode.

Fluidized Bed Bioreactors

Solid fluid contact may be achieved in these types of bioreactors. The fluid is pumped from the bottom from perforations and the solid particles are taken in the reactor. When the fluid velocity is increased the solid particles will get fluidized. The gravitational force, acting down on the particle, is counter acted by the buoyancy forces by the fluid going up. Mostly the fluidized bed bioreactors are used in wastewater treatment systems.

Stability of Bioreactors

The designing and control of bioreactor systems involves the stability of the bioreactor. When a steady state is locally stable, the system concentration will return to the steady state even after a small distance moves those concentrations slightly away from the reference steady state. Normally the stable bioreactor system is governed by the parameters like cell growth and cell morphology.

7.3 Bioreactor Dynamics

The dynamic characteristics of the bioreactors are considered here. The dynamics of C.S.T.R are considered which may be further extrapolated to other type's reactors. Initially, for unsteady state reactor performance the equations are developed. The dynamic model proposed by Bailey and Ollis (1986) for unsteady state is d/dt (total amount in the reactor) = rate of addition to reactor- rate of removal from reactor + rate of formation within the reactor.

So in a C.S.T.R,

$$\frac{d(Vc_i)}{dt} = v(c_{io} - c_i) + vr_{fi} \qquad \qquad .....(1)$$

Where

V = volume of the reactor.

v = volumetric feed rate.

r_{fi} = Rate of formation of component 'i' in the reactor.

c_i = Local concentration of component 'i' in the reactor.

c_{io} = Concentration of component 'i' in the feed.

Equation (1), dividing by v and where D (dilution rate) = v/V, becomes,

$$\frac{d(c_i)}{dt} = D(c_{io} - c_i) + r_{fi} \qquad \qquad .....(2)$$

In the above material balance, it is assumed that the feed stream and reactor contents have equal density and inlet and out-let volumetric flow rates are equal. The unsteady state material balance will be the starting point for reactor dynamics characterization. The dynamic characteristics of bioreactor are a function of concentration changes fc_i and function of parameter changes.

Now we will consider the situation where single substrate limits the growth cells and the performance of a bioreactor.

Upon application of unsteady state mass balance to growth of cells and substrate consumption and by using Monod expression, gives.

$$\frac{dx}{dt} = D(x_o - x) + \frac{\mu_{max} s_x}{k_s + s} \qquad \qquad .....(3)$$

$$\frac{ds}{dt} = D(s_o - s) - \frac{1}{y_{x/s}} \frac{\mu_{max} s_x x}{x + k_s} \qquad \qquad .....(4)$$

In the equation (3), for sterile feed $x_o = 0$. There are two possible states.

(i) non – initial steady state solution.

$$X_{ss} = Y_{x/s}(s_R - s_{ss})$$

$$s_{SS} = \frac{k_s (D)}{\mu_{MAX} - D}$$

OR

(ii) Washout steady state situation.

$$X = 0, \; s = s_o$$

The Monod-Chemostat study results for steady state washout condition are given as

For

1. $D > \dfrac{\mu_{MAX} \, s_O}{K_S + s_O}$ - Chemostat performance is stable.

2. $D < \dfrac{\mu_{MAX} \, s_O}{K_S + s_O}$ - Chemostat performance is unstable.

7.4   Biomass Production

Considering the cell balance across the reactor,

We have,

$$\frac{dx}{dt} = D(x_o - x) + \frac{\mu_{max} \, sx}{k_s + s}$$

At steady state condition, dx/dt=0

So,

$$\left[\frac{\mu_{MAX} \, s}{K_S + s} \right] x + D \, x_o = 0 \qquad \qquad .....(5)$$

The balance on substrate gives

$$\frac{ds}{dt} = D(s_o - s) - \frac{1}{y_{x/s}} \frac{\mu_{max} \, s \, x}{k_s + s} \qquad \qquad .....(6)$$

At steady state ds/dt = 0

$$0 = D(s_o - s) - \frac{1}{y_{x/s}} \frac{\mu_{max} \, s \, x}{k_s + s} \qquad \qquad .....(7)$$

Equations (5) and (7) are called as Monod Chemostat model.

For a sterile feed $x_o = 0$, Equations (5) and (7) can be solved to yield equation.

$$X_{ss} = Y_{x/s}(s_R - s_s) \qquad \qquad .....(8)$$

$$s_{ss} = \frac{K_s D}{\mu_{max} - D}$$

Hence the steady state condition depends on dilution rate 'D'. For very slow flows at a given volume when dilution rate tends to zero S_{ss} tends to zero. The cell mass concentration X_{ss} will be equal to $= Y_{X/s} (S_R)$

It is seen that with the increase of 'D', s initially increase linearly (Fig 7.3). After a point 'P' in the Figure, since increase rapidly as dilution rate approaches μ_{max} and as s_{ss} become very high, X_{ss} and this condition is known as 'washout'.

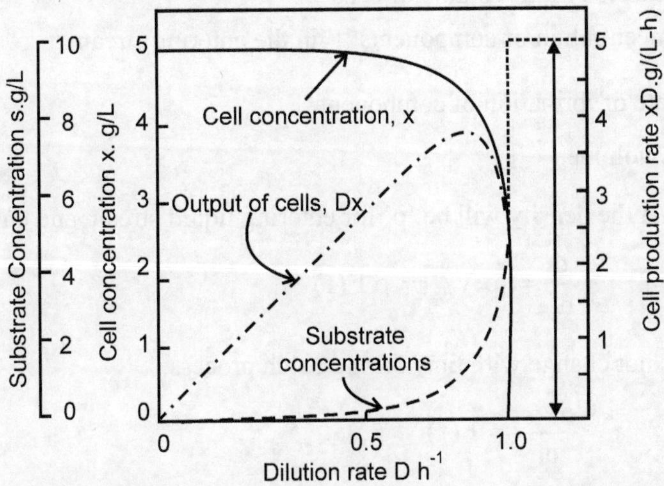

Fig. 7.3 Dependence of effluent substrate concentration s, cell concentration x, and cell production rate xD on continuous culture dilution rate D (Monod Chemostat)

So D_{max} becomes

$$D_{max} = \frac{\mu_{max}\, s_R}{K_s + s_R} \qquad \qquad .....(9)$$

Also the cell output i.e., x. D will be increased rapidly till the point P and decelerating sharply. Bailey and Ollis (1986) predict maximum value of cell production to avoid the region of large sensitivity.

7.5 Batch Reactors

In biochemical processes, it is always essential to add liquid streams to batch reactor during the process. Mostly, liquid streams such as nutrients, precursors, inducers etc; are added so as to minimize catabolic repression.

The liquid streams may be added in the process to extend stationary phase and to obtain additional products by adding nutrients. The culture volume altered when a liquid feed stream is added to the reactor, and described in equations.

$$\frac{d}{dt}[V_R.C_i] = V_R.r_{fi} + F(t) \qquad .....(10)$$

Where

$F(t)$ = volumetric flow rate of entering feed stream

C_i = stream moles i / unit volume of cells

$C_{if}(t)$ = concentration of components 'i' in the entering stream

r_{fi} = molar rate of formation of component 'l'

V_R = culture volume

It is assumed that the density will be 'ρ' for entering liquid stream and culture liquid.

$$\frac{d}{dt} = [\rho.V_R] = \rho.F(t) \qquad .....(11)$$

Now if 'ρ' does not change with time during batch process.

$$\frac{dV_R}{dt} = F(t) \qquad .....(12)$$

Carrying out differentiation of equation and upon dividing equation (10) by V_R and substituting in equ (12) given

$$\frac{dC_i}{dt} = \frac{F(t)}{V_R}(C_{if} - C_i) + r_{fi} \qquad .....(13)$$

Equation (12) and (13) are mass and component balance equation respectively.

Enzyme Catalyzed Reactions in CSTRS

The enzyme catalyzed reaction can be carried out in a variety of CSTRs (Fig 7.4). The various configurations are given below.

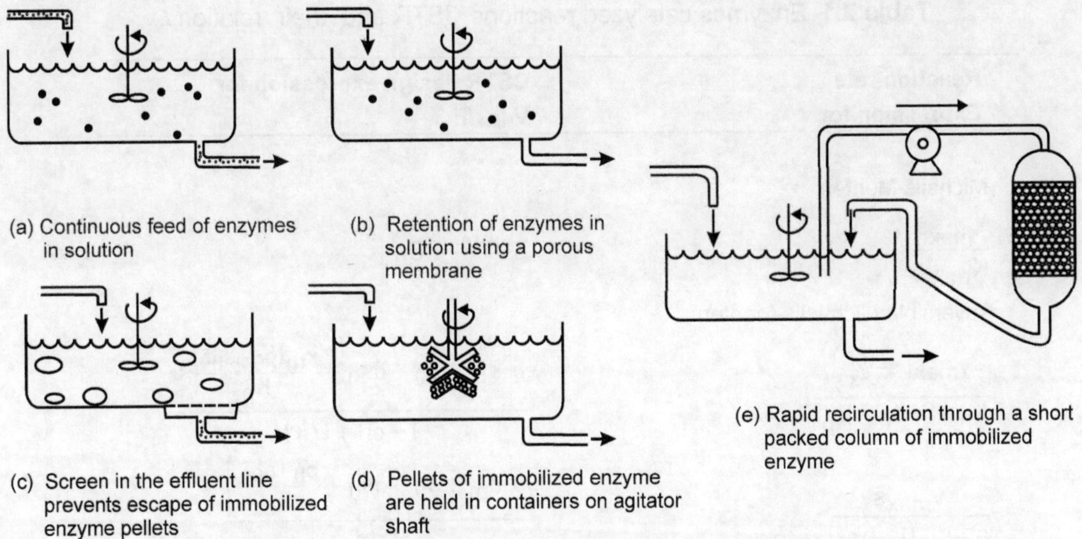

Fig. 7.4 CSTR systems for immobilized reactions.

In a simple system (1) enzymes are added to reactor and removed from the reactor through feed at out let streams. This is applied when the enzymes are cheaply available. Different configuration of CSTRs will be used for costly enzymes. (2) Ultra filtration membrane is used in the effluent stream which retains the enzymes in the reactor (3) A screen is filled in the effluent line which prevents the escape of enzymes. (4) Enzymes are attached to the agitated shafts in screen baskets. (5) Enzymes are fixed in packed column through which the reactor contents are circulated.

The main objective of the various reactor configuration is to maintain desired enzymes concentration inside the reactor.

Applying the basic principles of material balances, substrate and product concentration could be estimated.

Considering the basic equation

$$S \ \rightarrow \ P$$

If 1 mol of product P is for each mole of substrate 'S' and the feed (S_0, P_0) and effluent concentration (S, P) are depicted as,

$$S_0 - S = P - P_0 \qquad \qquad .....(14)$$

The substrate mass balance considering reaction rate expression $v(s,p)$ can be written as ,

$$F\,(s_0 - s) - V_R v\,(s, P_0 + s_0 - s) = 0 \qquad .....(15)$$

The solution to this equations are given in Table (7.1).

Table 7.1 Enzymes catalyzed reactions CSTR and their relations.

Reaction-rate Expression for v	CSTR design expression for V_{vmax}/F
Michalis-Menten: $\dfrac{v_{max}s}{K_m+s}$	$\delta\left(\dfrac{k_m}{1-\delta}+S_0\right)$
Reversible Michaelis-Menten: $\dfrac{v_{max}\left(s-\frac{p}{k}\right)}{K_m+S+K_{mp}/K_p}$	$\dfrac{\delta\left[k_m+s_0-\delta s_0+\dfrac{k_m(p_0+s_0\delta)}{K_p}\right]}{1-\delta(1+1/k)}$
$\dfrac{v_{max}s}{a+k_m(1+pk_1)}$	$\dfrac{\delta\left[k_m+s_0-\delta s_0+\dfrac{p_0+s_0\delta}{k_1}\right]}{1-\delta}$
Substrate inhibition : $\dfrac{v_{max}}{1+\frac{k_m}{s}+\frac{s}{k_1}}$	$\delta s_0\left[1+\dfrac{k_m}{s_0(1-\delta)}+\dfrac{(1-\delta)s_0}{k_1}\right]$

CSTR with Cell Cycle and Cell Growth

The CSTR with a separator at out let stream and recycle from separator to feed will enhance biomass in reactor and results in increase in product yield. This shown in the (Figure 7.5).

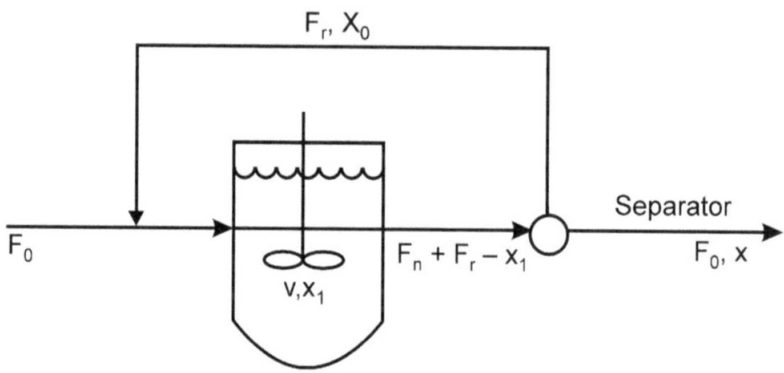

Fig. 7.5 CSTR with recycle.

Where

F_0, F_r = feed and recycle volumetric flow rates.

X_1, x_0, x = reactor, recycle stream and products streams biomass concentrations.

The steady state biomass considerations equation for recycle is given as,

$$F_r x_0 + \mu x_1 V_R - (F_0 + F_r)x_1 = 0 \qquad .....(16)$$

Where

$$A = F_r / F_0$$

$$b = x_0 / x_1$$

The dilution rate is separated as,

$$D = \frac{\mu}{1 - a(b - 1)} \quad\quad .....(17)$$

The microorganism in recycle liquid is more concentrated than the microorganisms in the effluent and b>1.

It is known from the eqn (17) that dilution rate D is larger than organism specific growth rate μ.

It can be visualized that reactor with recycle can process more feed material than the reactor with-out recycle.

The substrate balance upon assumption of constant yield factor is

$$D(S_0 - S) - \frac{\mu x_1}{Y} = 0 \quad\quad .....(18)$$

After combining eqn (17) with (18) the biomass production rate per unit reactor volume μ_x, is

$$\mu x_1 = \frac{\mu Y (S_0 - S)}{1 - a(b - 1)} \quad\quad .....(19)$$

Recycle production rate is greater by $[1-a(b-1)]^{-1}$ than non-recycle production rate.

During cell cultivation, it is observed that the dilution rates are high without washout, than observed ideal CSTR. This is possible because of growth of cells on the wall.

The organism will be seen as solid films on the reactor wall.

Now considering cells on the film at the reactor, may be above reactor liquid level as x_f. The cells reproduced on solid films at the reactor could fall back into the reactor.

The steady state continuous reactor mass balance will be in the form,

$$Dx = \mu x + \mu_f x_f \quad\quad .....(20)$$

$$D(s_0 - s) = \frac{1}{\tau}\mu x + \frac{1}{Y_f}\mu_f x_f \quad\quad .....(21)$$

Where μ_f = specific growth in the film

$\quad\quad Y_f$ = yield factor in the film

This may offer from μ and Y for various reasons including diffusion reactions.

Ideal Plug Flow Reactors

The plug flow is expected when fluid moves in large channel or in pipe with high Reynolds number i.e., more than 2100 in a pipe and there is no variation of axial velocity over the cross-section. In contrast to ideally mixed bioreactor, there is no mixing in, plug flow reactor (Fig 7.6). The length to diameter ratio is high in plug flow reactor. Substrate enters from one side and leaves process other side while the reaction takes place by microbial population within the reactor. The concentration of substrate decreases as it flows along the bio reactor since there is no mixing in the bioreactor. However the substrate concentration will be uniform in the completely mixed bioreactor. The packed bed immobilized bioreactor behave like plug flow bioreactor.

It is assumed that plug flow prevails in the bioreactor; the mass balance can be made by differential section approach in plug flow reactor.

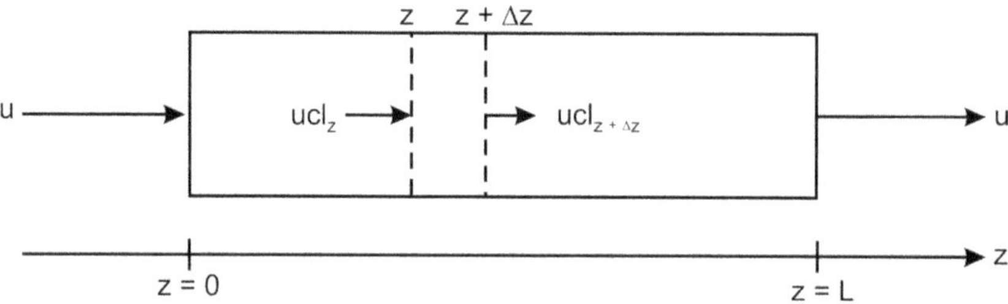

Fig. 7.6 Ideal plug flow reactor.

The mass balance on then section can be taken as,

$$A\upsilon c \big|_z - \ A\upsilon c \big|_{z+\Delta z} + A\Delta_z r_{fc} \big|_z = 0 \qquad \qquad .....(22)$$

Where

r_{fc} = rate of formation of species 'c' = amount / unit volume . time

Rearranging and dividing by A Δz we get,

$$\frac{Vc \big|_{z+\Delta z} - V \big|_z}{\Delta z} = r_{fc} \qquad \qquad .....(23)$$

Taking limits

$$\frac{d}{dz}(vc) = r_{fi} \qquad \qquad .....(24)$$

Upon assuming no change in density during reaction, the axial velocity will be constant and equation (24) changes to,

$$\frac{dc}{dz} = r_{fc} \qquad \qquad .....(25)$$

The time required to move one fluid section from entry point of reactor to position of z is z/v

Is

$$t = \frac{z}{v} \qquad \qquad .....(26)$$

The mass balance equation (25) can be written as,

$$\frac{dc}{dt} = r_{fc} \qquad \qquad .....(27)$$

Equation (27) is same as the mass balance equation of batch reactor. The plug flow with constant velocity and each segment moves with no interaction, with neighboring segments. The system is totally segregated and each segment behaves as a batch bioreactor. Also if the initial feed to batch reactor and feed at entry point of plug flow reactor are same and L/v the residence time of PFR is same as batch time, then the tube effluent will be same as batch reactor effluent.

So the boundary condition for this will be

$$C\big|_{z=0} = C_0 \qquad \qquad .....(28)$$

Where

Z = 0 inlet to reactor

C_0 = concentration of feed at inlet

If Monod Chemostat is applicable to PFR,

The mass balance on cells and substrate would be,

$$\frac{dx}{dt} = \frac{\mu_{max}\, x\, s}{s + k_s} \qquad \qquad .....(29)$$

$$\frac{ds}{dt} = -\frac{1}{Y}\frac{\mu_{max}\, x\, s}{s + k_s} \qquad \qquad .....(30)$$

Applying initial conditions

$$x(0) = x_0$$

$$s(0) = s_0 \qquad \qquad .....(31)$$

Equation (29) & (30) upon computations and substrate and cell concentration, can be stoichiometrically written as,

$$X + Y_s = x_s + Y_{so} \qquad \qquad \dots\dots(32)$$

Using eqn. (32) and expression x in terms of s, and substituting in eqn. (29), we get

$$\frac{ds}{dt} = \frac{\mu_{max\ s}}{Y} \frac{\left[x_0 + Y(s_0 - s)\right]s}{s + k_s} \qquad \qquad \dots\dots(33)$$

Integration of this and applying initial conditions, we get,

$$x_0 + Y(s_0 + K_s) \ln \frac{x_0 + Y(s - s_0)}{x_0} - K_s Y \ln \frac{s}{s_0} = \mu_{max} t(x_0 + Y_{so}) \ \dots\dots(34)$$

The outlet substrate concentration will be 's' value with respect to t=L/v and then x is formed with eqn (32)

7.6   Residence Time Distribution Concept

What would be the fate of small segment of fluid which enters the continuous flow bioreactor? This small segment of fluid will be further reduced to smaller particle due to mixing in the reactor and disperse throughout the bioreactor. Ofcourse, some part of these smaller particles will reach effluent stream. Nevertheless the remaining particles would be spending various times before actually reaching the effluent stream. Hence the effluent stream would be the mixture of various particles that have resided for different length of time inside the bioreactor. Now the determination of residence times of the particles at the effluent stream will give the information about mixing and flow patterns in the bioreactor.

Stimulus – response studies are conducted by introducing tracer as stimulus to know the extent of stay of elements inside the bioreactor. The tracer is selected such that, it has the following characteristics

1.   It will not react with the reactants.

2.   It will be easily detected

3.   It will not disturb the flow pattern

Tracers like salts, acids or base or hydrocarbon gas are introduced either as step input or as pulse input.

The stimulus and response features are depicted in the (Figure 7.7)

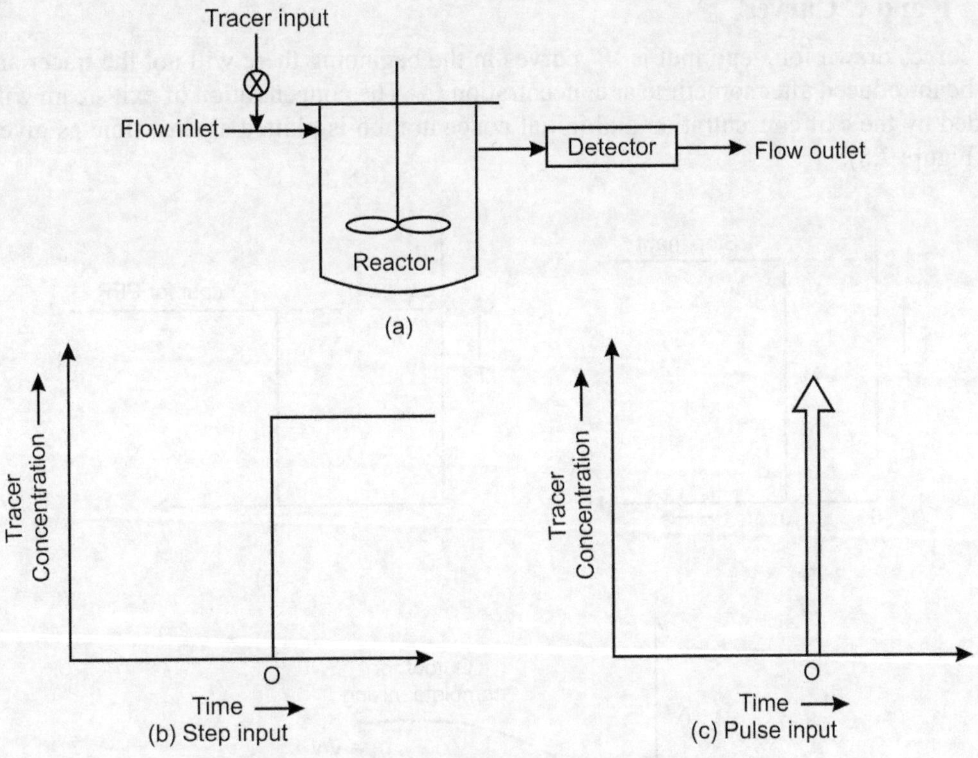

Fig. 7.7 Stimulus response studies on CSTR.

Stimulus – Response

The quality of the tracer introduced is 'Q'. The tracer before leaving the reactor, due to non-ideality in flow resides for different time periods in the reactor. Obviously this indicates that each fraction of tracer will spend different time inside the reactor. A fluid element may spend time between t and t + dt. 'E' the exist age distribution is known as the study of the all the fraction of the tracer or fluid which spends different times in the reactor.

Now E for all the fractions which spend between o & t is $\int\limits_{0}^{t} Edt$

Also the fractions which spent between t and ∞ is given by $\int\limits_{t}^{\infty} Edt$

$$\int\limits_{0}^{t} Edt + \int\limits_{t}^{\infty} Edt = \int\limits_{0}^{\infty} Edt = 1 \qquad \qquad .....(35)$$

The total of the fraction is 1

7.7   F and C Curves

The curves drawn for step input is 'F' curves in the beginning there will not the tracer and it will be introduced after sometime at concentration C_0. The concentration of exit steam will be divided by the exit concentration and initial concentration is plotted against time as given in the (Figure 7.8).

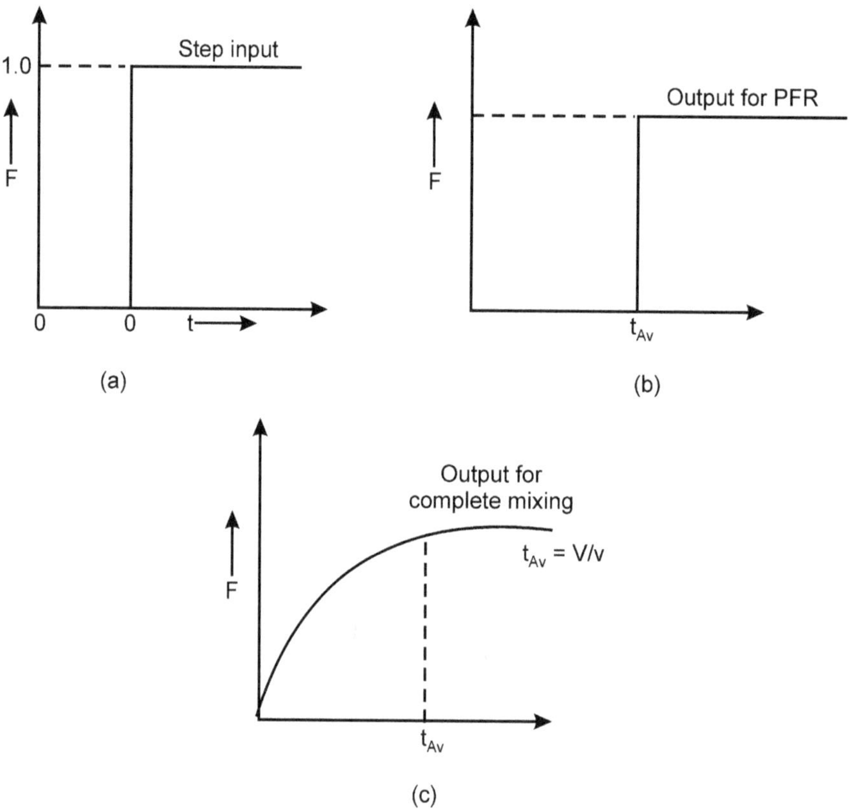

Fig. 7.8 F-curves (a) step input (b) Out put for PFR (c) out put for CSTR

The exit age distribution will be

$$\int_0^t Edt \qquad \qquad .....(36)$$

The output responses for the step input to PFR will give out response after t_{av}, the space time. So the first fraction of element comes out after t_{av}, and response is similar to input in PFR. Whereas the contents are mixed completely in CSTR and the first element come out immediately after injection. The concentration slowly builds up over a period of time. (as Fig (7.8c))

The curve drawn per pulse input is known as curve 'C'

$$C = c/Q \qquad \qquad .....(37)$$

C = concentration of tracer at any given time as response

$$\text{So } Q = \int_0^\infty c\,dt \qquad \qquad .....(38)$$

The response curve per pulse input will be equal to exist age distribution C = E

$$\int_0^\infty c\,dt = \int_0^\infty \frac{c}{Q}\,dt \qquad \qquad .....(39)$$

Thus,

the relationship of three curves is

$$E = C = \frac{dF}{dt} \qquad \qquad .....(40)$$

The C curves are drawn the Figure (7.9).

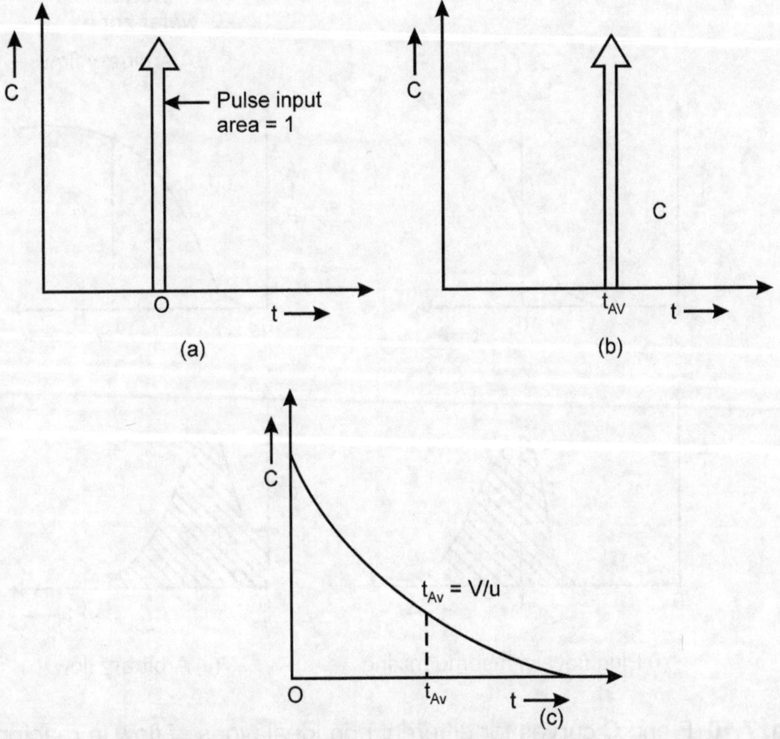

Fig. 7.9 C – curves showing (a) pulse input (b) PFR out put (c) CSTR out put.

When pulse input is given to PFR, the response come out in the focus of pulse after t_{av} space time (Fig 7.9b). The input pulse and out put response are similar but for the response visible after t_{av}. This indicates that three will not be any tracer in the response before t_{av}. Since the contents are uniformly mixed in CSTR trace will have uniform concentration. So the maximum concentration of tracer in response curve is observed at the beginning from the zero time and it reduces slowly (Fig 7.9c).

The F and C curves for F and C are shown in Figure 7.10.

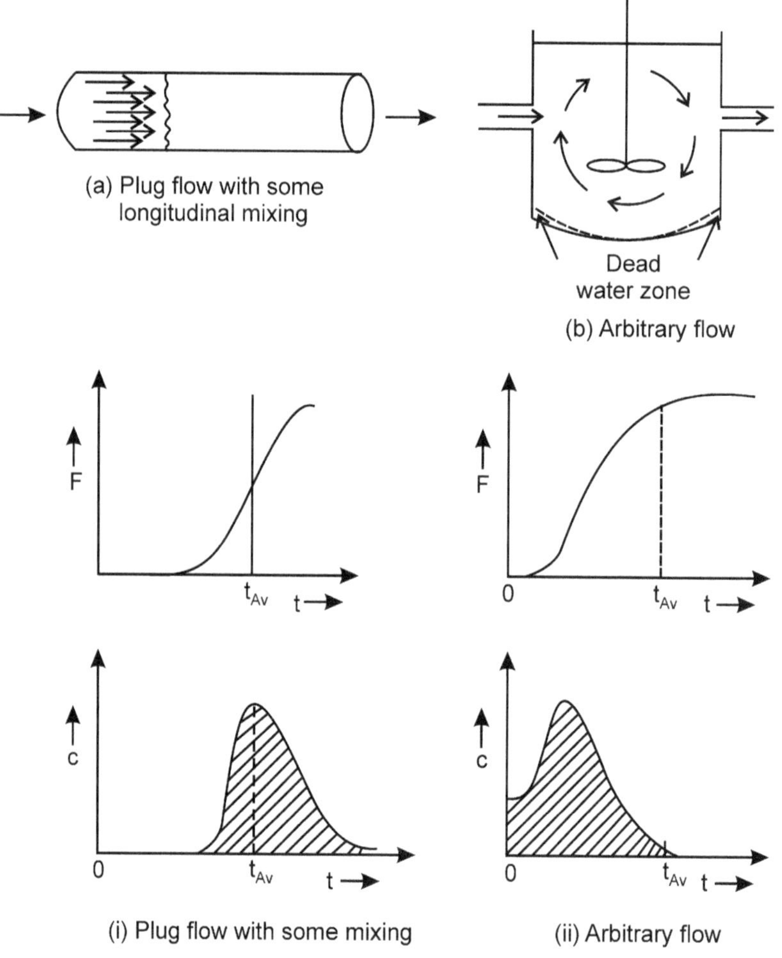

Fig. 7.10 F and C curves for different non-ideal types of flow in reactors.

The non ideal flow response, curves, patterns will be in between the curve patterns of PFR and CSTR and the response starts in between o and t_{Av} (shown in the Fig. 7.10)

7.8   Mean Residence Time

Various elements will make per different times in the reactor. So it is essential to evaluate mean residence time. It is the ratio of the total of fractions of tracer material multiplied by its residence time to the total quantity of tracer introduced to the reactor.

$$t_{AV} = \frac{\int_0^\infty t^c dt}{\int_0^\infty cdt} = \frac{\sum_{i=1}^\infty t_i C_i \Delta t_i}{\sum_{i=1}^\infty C_i \Delta t_i} \qquad \qquad .....(41)$$

The distribution of residence time over average residues time (t_{Av}) is various

$$\sigma^2 = \frac{\int_0^\infty (t - t_{Av})^2 C dt}{\int_0^\infty cdt} = \frac{\sum_{i=1}^\infty t_i^2 C_i \Delta t_i}{\sum_{i=1}^\infty C_i \Delta t_i} - t_{Av}^2 \qquad \qquad .....(42)$$

Conversions

The conversion of the reactant can be estimated by measuring the exit age distribution of the given tracer element. The exit age distribution of the tracer and the reactant are assumed to be the same. This means that the outlet concentration distribution of the reactant is same as tracer age distribution (E).

The equation for the unconverted reactant can be given by

$$C_{A,\,Av} = \int_0^\infty (C_{Areactatt})(Edt) \qquad \qquad .....(43)$$

Where $\dfrac{C_{Areactent}}{C_{A0}} = e^{-k_1 t}$ for the first law reaction kinetics

Equ (43) may be written as,

$$\frac{C_{A,AV}}{C_{A0}} = \sum e^{-K_1 t_i} (E_i \Delta t) \qquad \qquad .....(44)$$

The out-let average concentration of the reactant

$$= \Sigma \text{ (concentration of the reactant)} \quad - \quad \text{(Fraction of exit steam, in}$$
$$\text{Between t and t + dt)} \qquad \qquad \text{overall stream between t,}$$
$$\text{and t + dt)}$$

$$.....(45)$$

7.9   Models

The models used for non-ideal behaviuor are

(a)   Dispersion model   (b)   Tanks in series model

Dispersion Model

This model basically depends on the certain assumptions like (a) no short circuiting, (b) no by passing and (c) no stagnant zones. Dispersion model will not be appropriate for CSTR. However the non-ideality will be due to intermixing or back mixing in plug flow reactors. Normally certain amount is expected between different layers due to flow disturbances. This sort of intermixing by molecular movement is known as molecular diffusion as give by.

$$\frac{\partial c}{\partial t} = D \frac{\partial^2 c}{\partial x^2} \qquad .....(46)$$

Where D = diffusivity coefficient

Similarly back mixing or intermixing can be expressed as,

$$\frac{\partial c}{\partial t} = D_e \frac{\partial^2 c}{\partial x^2} \qquad .....(47)$$

Where D_e = axial dispersion coefficient

The two dimensional parameters for length and time could be,

$$Z = \frac{X}{L} \qquad .....(48)$$

$$\Theta = t/t_{Av} = tu/L \qquad .....(49)$$

Where z = dimensional less length (in PFR)

Θ = dimensional less time (Average residence time PFR) substituting values of 48, 49 in equation 47

We have

$$\frac{\partial C}{\partial \theta} = \left(\frac{D_e}{uL}\right) \frac{\partial^2 C}{\partial z^2} - \frac{\partial C}{\partial z} \qquad .....(50)$$

Where $\dfrac{D_e}{uL}$ = Vessel dispersion number. This gives extent of axial dispersion

If $D_e/UL = 0$ no dispersion - so flow is, plug flow

If $D_e/UL = \infty$ Dispersion is large – so flow is mixed flow

$1/(D_e/UL)$ is dimensional less number i.e., Pecklet number (N_{Pe})

$$N_{Pe} = \frac{\text{Convection rate of transport}}{\text{Diffusion or dispersion rate of transfer}}$$

$$N_{Pe} = \frac{uL}{D_e} \qquad\qquad .....(51)$$

Another dimensional less number

$$N_{De} = \frac{\text{Consumption of rectant A (by clinical reaction)}}{\text{Transport of reactant A (by convection)}}$$

$$= K_1\, t_{Av} \text{ (first order reaction)}$$

Dimension less variance

$$\sigma_\theta^2 = \frac{\sigma^2}{tAv^2} \qquad\qquad .....(53)$$

the relation ship between σ_θ^2 and N_{Re} is given as

$$\sigma_\theta^2 = \frac{2}{N_{Pe}} - \frac{2}{N_{Pe}^2} - (1 - e^{-N_{Pe}}) \qquad\qquad .....(54)$$

Estimation of Conversion (Dispersion)

Mass balance equation at differential length of a tubular reactor will be used per calculating conversion with dispersion.

The first reaction kinetics can be given as

$$\frac{\partial^2 C_A}{\partial x^2} - \frac{\mu}{D_e}\frac{dC_A}{dx^2} - \frac{k_1 C_A}{D_e} = 0 \qquad\qquad .....(55)$$

So, $z = x/L$ and multiplying by (D_e/v), we get,

$$\frac{D_e}{uL}\frac{\partial^2 C_A}{L\partial t^2} - \frac{1}{L}\frac{dC_A}{dx^2} - \frac{k_1 C_A}{u} = 0$$

Put $L/v = t_{Av}$ after multiplying by L

$$\frac{D_e}{uL}\frac{\partial^2 C_A}{\partial z^2} - \frac{dC_A}{dz^2} - K_1 C_A t_{Av} = 0 \qquad\qquad .....(56)$$

The above equ (56) is similar to equation (50)

Where θ is replaced by t_{Av}/t and $\dfrac{dC_A}{dt}$ is conversion replaced by

$K_1 C_A$ for first law kinetics

Equation (56) can be written as

$$\frac{D_e}{uL}\frac{\partial^2 X_A}{\partial z^2} + K_1 t_{A1}(1 - X_A) = 0 \qquad\qquad .....(57)$$

The analytical solution to above equation is given by,

$$\frac{C_A}{C_{Ao}} = (1 - X_A)$$

$$\frac{C_A}{C_{Ao}} = (1 - X_A)$$

$$= \frac{49\exp(N_{Pe}/2)}{(1+q)\exp(N_{Pe}q/2) - (1-q)^2\exp(-N_{Pe}q/2)} \qquad .....(58)$$

Where $q = (1 + 4(N_{Pa}/N_{Pe}))^{1/2}$

Tanks in Series

CSTR equal tanks connected in series (Fig 7.11) the exit age distribution relation to N (number of tanks)

$$t_{Av}E = (\frac{t}{t_{Av}})^{N-1}\frac{1}{N-1}e^{-t/t_i Av} \qquad .....(59)$$

Where

t_{cAv} = Mean residence time of ith tank for 'N' tanks

$\theta_i = t/t_{c,Av}$

This

$$\theta = \frac{t}{t_{Av}} = \frac{t}{N_{t_i,Av}} \qquad .....(60)$$

$$\sigma^2{}_\theta = \frac{\sigma^2}{t_{Av}{}^2} = \frac{1}{N} \qquad .....(61)$$

$$E_\theta = t_{Av}\,E \qquad .....(62)$$

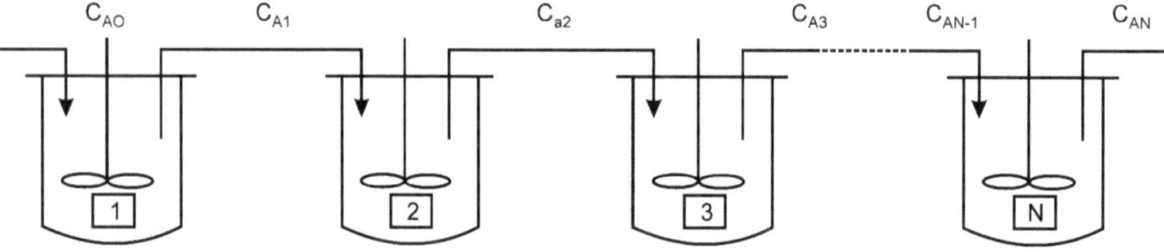

Fig. 7.11 N- tanks connected in series.

Estimation of conversion : (Tanks in series)

The conversion in CSTR – first order reaction is given as,

$$\frac{C_A}{C_{A0}} = \frac{1}{1 + k_1 t_{Av}}$$

$$\frac{C_{A0}}{C_A} = 1 + k_1 t_{Av} \qquad \dots(63)$$

As shown in Figure (7.11) in N-tanks connected in series, the outlet of first reactor would be inlist of second reactor and like that,

The first reaction conversion is

$$\frac{C_{A0}}{C_{A1}} = 1 + k_1 t_{1,Av} \qquad \dots(64)$$

Second reactor, conversion is

$$\frac{C_{A1}}{C_{A2}} = 1 + k_1 t_{2,Av} \qquad \dots(65)$$

N^{th} reaction conversion

$$\frac{C_{AN-1}}{C_{AN}} = 1 + k_1 t_{N,Av} \qquad \dots(66)$$

$$\frac{C_{A0}}{C_{AN}} = \frac{C_{A0}}{C_{A1}} \times \frac{C_{A1}}{C_{A2}} \times \dots\dots\dots \times \frac{C_{AN-1}}{C_{AN}} \qquad \dots(67)$$

$C_A = C_{AN}$ = final out let concentration

So equation (67) becomes,

$$\frac{C_{A0}}{C_A} = 1 + k_1 t_{1,Av})(1 + k_1 t_{2,Av})\dots\dots(1 + k_1 t_{N,Av}) \qquad \dots(68)$$

Since flow rate are same in equal volume reactions, and assumptions dusty change in negligible,

$$t_{1,Av} = t_{2\,Av} = t_{N,Av} = \frac{V}{v} \qquad \dots(69)$$

where

    v = volume of each rector

    υ = volumetric flow rate

so over all space time will be the total of individual space times

$$t_{Av} = \sum_{i=1}^{N} t_{i,Av} = N t_{i,A_0} \qquad \dots(70)$$

combing (68) and (70) equation, we have

$$\frac{C_{A0}}{C_A} = \{1 + k_1(t_{Av}/N)\}^N \qquad \qquad .....(71)$$

$$\frac{C_{A0}}{C_A} = 1 - X_A = \frac{1}{\{1 + k_1(t_{Av}/N)\}^N} \qquad \qquad .....(72)$$

$$X_A = 1 - \frac{1}{\{1 + k_1(t_{Av}/N)\}^N} \qquad \qquad .....(73)$$

7.10 Design of Bioreactor

Microbial products of commercial importance can be produced by means of various bioreactor environment like microbial, enzymatic, animal or plant or fungal systems. But the main compound of all such processes is bioreactor. So the design and operation of bioreactors plays an important role in any of the biological process. These bioreactors may operate with microorganisms (1) microbial bioreactor (2) Enzyme reactors (cell free).

The microbial reactors have been operated with microbial cells and enzyme reactors have been operated with enzymes which are free of microbial cells.

In a typical microbial process the number of microorganism will be increased during a biochemical reaction and stirred tank reactors are generally used. In enzyme catalyzed reactions, the concentration of enzyme will remain same either stirred tank reactor or plug flow reactions would be suitable. The concentration time profiles are given in Figure (7.12)

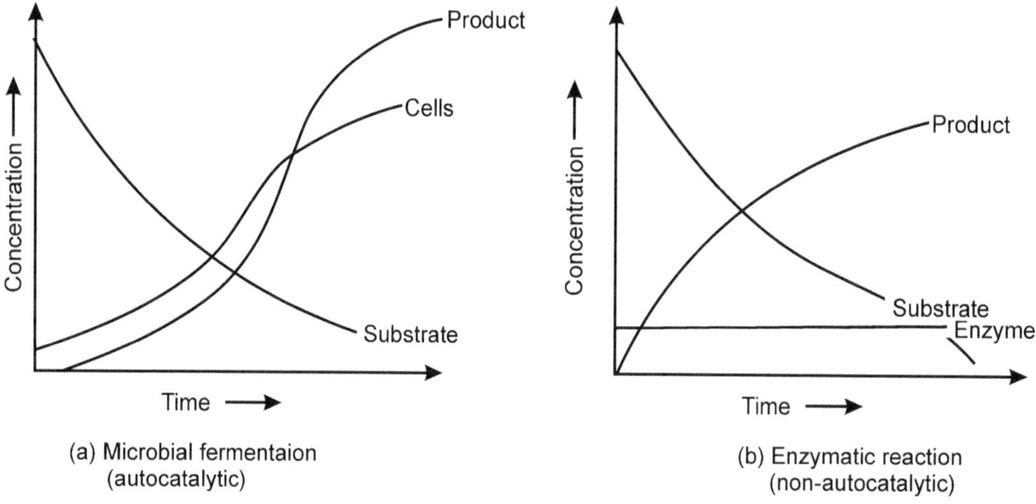

Fig. 7.12 Concentration Vs time diagram for (a) microbial reaction (b) enzyme catalyzed reactions.

The bioreactor may be classified as per in process conditions as 1. aerobic 2. anaerobic 3. immobilized 4. Solid state.

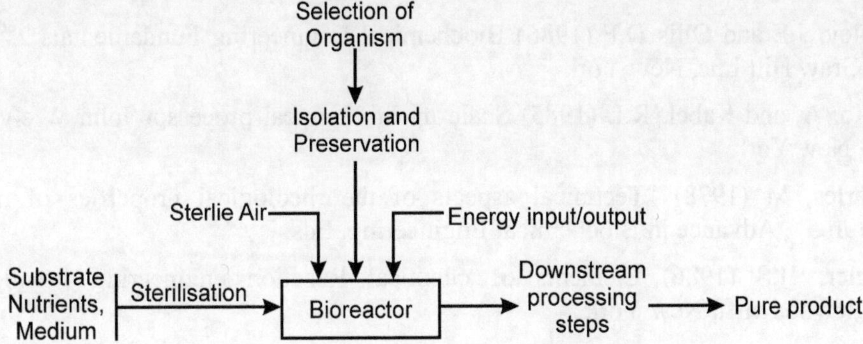

Fig. 7.13 Steps of a typical biological process.

In a biological process, (Fig 7.13) the operation of bioreactors depends various factors. Some of them are discussed here. Such factors include

1. Cell growth and cell concentration

2. Heating and cooling systems

3. Sterile conditions maintained

4. Agitation and mixing systems

5. Optimal shear rate maintenance

6. Filamentous growth associated readings

7. Very low growth rates of microorganism in large reactor and residence times. Mass transfer aspects are considered while designing bioreactors. Most of the microbial reactions are aerobic in nature and gas liquid transfer is often found in biological systems.

The design equation for such gas liquid system is,

$$N_A = k_{La}(C_{AL}^* - C_{AL}) \qquad\qquad .....(74)$$

N_A = mass transfer rate (kg/m^{3}s)

k_L = liquid side mass transfer coefficient (m/s)

a = equilibrium area (m^2/m^3)

C_{AL}^* = equilibrium concentration of the gaseous component A in the liquid phase (kg/m^3).

The mass transfer coefficient K_{La} values depend on different factors like agitation rate, bubbles size, gas sparging, system, gas velocity, cell viscosity, temperature, and viscosity of the fermentation broth.

References

1. Aiba, S, Humphrey A.E and Millis N.F (1973) Biochemical Reference : Engineering , Academic press , New York

2. Bailey, J.E and Ollis D.F (1986) Biochemical Engineering Fundamentals 2nd edition, McGraw Hill Lnc, New York

3. Bisio, A, and Kabel, R.L (1985) Scale up of chemical processor John Wiely & Sons lnc, New York

4. Charles, M (1978) "Technical aspects of the rheological properties of microbial cultures ", Advance in Biochemical Engineering, Eds.

5. Fogler, H.S (1986) Elements of chemical Reaction engineering Prentice Hall, Englewood cliff, New York

6. Kunni, D., and Leven spiel, O. (1991) Fluidization engineering, 2nd Edn., Butter Worth –Heinemann, Boston MA.

7. Shuler M.L and Kargi F (2003) Bioprocess engineering, Basic concepts 2nd Edn., Prentice- Hall India, New Delhi.

Review Questions

1. Write briefly about bioreactor dynamics.

2. Describe batch bioreactors.

3. Explain enzyme catalyzed reactions in CSTRs.

4. Write about CSTR with cell recycle.

5. Describe Residence time distribution concept.

6. Explain dispersion model and Tanks in series models.

7. Briefly describe about design of bioreactors.

CHAPTER 8

Medium and Bioreactor Sterilization

The medium which is used for a biochemical process must be free from any microorganism. Depending on the product desired specific microorganism will be inoculated to the production medium. So, it is mandatory to make the medium devoid of microorganism, prior to inoculation with a particular strain. Various methods may be adopted to remove microorganisms, like filtration, flotation and centrifugation. The other methods may include heat destruction, electromagnetic waves or chemical agents. Mechanical means of destruction like mechanical abrasion may be useful at small scale but this does not give satisfactory results at large-scale. At the laboratory scale certain methods like, X-rays, β-rays, u-v rays and sonic irradiations may be useful and they are not useful at commercial scale. However moist heat could be a better alternative to remove microorganisms from the process medium.

Continuous sterilization of media is gaining importance over batch sterilization due to various advantages like

1. Increase of productivity due to less damage to media constituents
2. Easy adoption to automatic control
3. Small utilization of process steam
4. Better quality control

The viscosity of the media must be low and foaming also must be controlled for a better operation of continuous sterilizer.

To understand the design and operation of sterilization equipment for sterilizing the media, it would be necessary to know the concept of thermal death of microorganism. The knowledge of kinetics of death of microorganism will be useful in the optimum design of sterilizers.

Thermal death studies are essential before actually subjecting the media for sterilization. These studies are required to find out the sterilization time, which depends on the nature of the media.

8.1 Thermal (Destruction) Death Microorganisms

The thermal death of microorganism is achieved by applying moist heat. The destruction of microorganism follows first order chemical reaction and can be expressed by :

$$\frac{dN}{dt} = -KN \qquad \qquad .....(1)$$

where N = the number of viable organism present

k = the reaction rate constant, mm^{-1}

t = sterilization time

The negative sign in the above equation represents, N decreases as t- increases.

Eqn (1) can be written as

$$\ln \frac{N_o}{N_t} \int_0^t Kdt$$

Another term decimal reduction time may be introduction here. It is the time of response of organism to heat, so that, the original number viable microorganisms will be reduced by one tenth. Integrating eqn (1)

$$\ln \frac{N_o}{N_t} \int_0^t Kdt$$

$$\frac{N}{N_o} = e^{-kt}$$

$$N = N_o e^{-Kt} \qquad \qquad .....(2)$$

By this definition of 'D"

$$\frac{N}{N_o} = \frac{1}{10} = e^{-kD}$$

$$D = 2.303 \,/K \qquad \qquad .....(3)$$

Bacterial spores are more resistance to heat than the vegetative cell. This indicates that the k values for the vegetative cells are more than the k- values for the spores. Probably dipicolinic acid present in spores may be responsible for their increased resistance to heat.

Fig (8.1) and Fig (8.2) represent typical death rate for bacterial spores and vegetative cells of bacteria Fig (8.1) Typical death rate for bacillus sp spores

N = no of viable species

N_o = original viable cells

Fig (8.2) Typical death rate E. coil

N = viable cells at

N_o= original values of viable cells

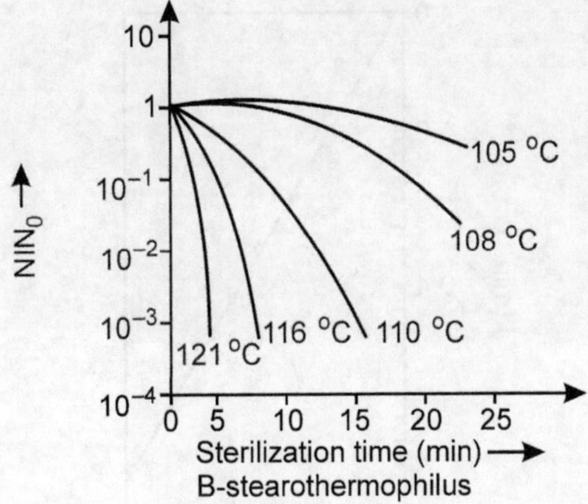

Fig. 8.1 Typical death rate data for Bacillus species spores.

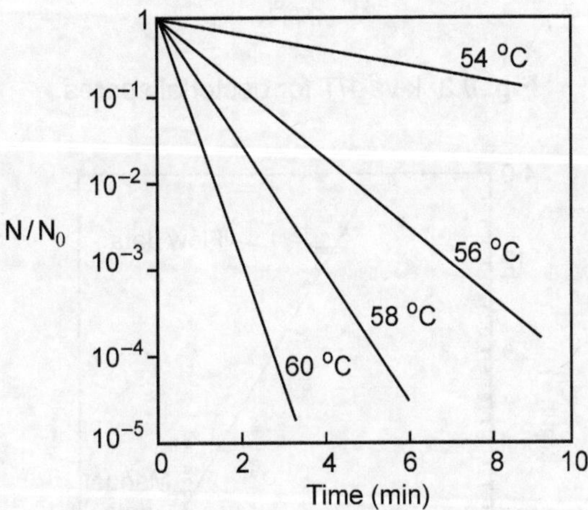

Fig. 8.2 Typical death rate for E.coli.

These plots were drawn for different temperatures. The two graphs indicate that bacterial spores are more resistance to heat then the vegetative cells of the bacteria. It has also been observed that the number viable spores increased immediately after exposure of heat and sharply dropped sterilization proceed further.

However, as the number of microorganisms surviving the heat sterilization is normally less than 1, probabilistic approach may be better considered rather then deterministic for the death rate of the microbial population.

Temperature Effect on Death Rate

The temperature effect on reaction rate constant 'k' is show ever in Fig. (8.3) for bacterial spores Fig. (8.4) for vegetative cells of bacteria.

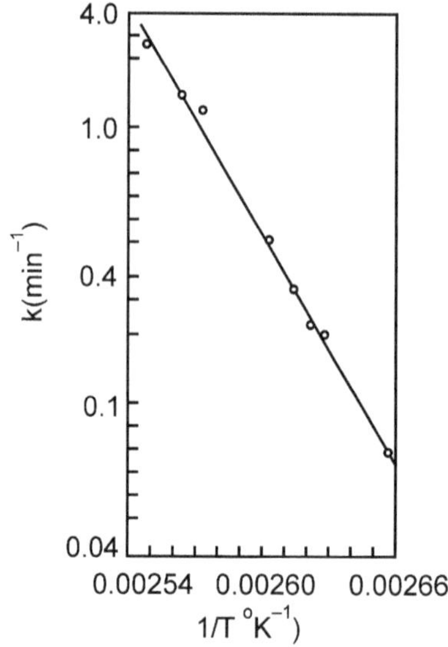

Fig. 8.3 k vs 1/T for bacterial spores.

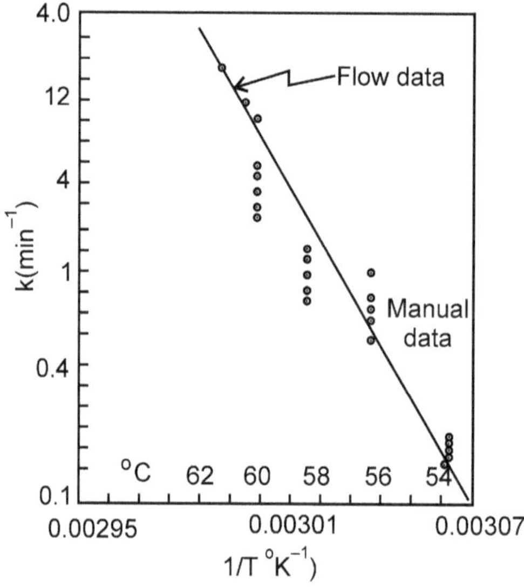

Fig. 8.4 k vs 1/T for vegetative cells of bacteria.

The K values plotted against 1/T i.e., reciprocal of absolute temperature

where   k = reactor rate constants, min^{-1}

T = absolute temperature ^{0}k

E = activation energy

where k = reactants constant, min^{-1}

   T = absolute temperature

   E = activation energy = kcal / g mole

The temperature effect on reaction rate constant can be expressed as Arrhenius equation

$$K = ά\ e^{-E/RT} \qquad \qquad .....(4)$$

   $α''$ = empirical constant

   T = absolute temperature $°k$

   E = activation energy kcal / g mole

   R = gas constant

As per the Eyring theory

$$K = g\ Temp\ (-\ \Delta H^*\ RT)\ exp\ (\Delta S^*/\ R) \qquad \qquad .....(5)$$

Where g = factor including (Boltzman constant and plank's constant)

   ΔH^* = heat of reactor of activation

   ΔS^* = entropy change of activation

As per Bigelen (1921) Q_{10} theory

$$D = α''\ exp\ (-β''\ T'') \qquad \qquad .....(6)$$

Where

   D = decimal reduction time

   $α''\ β''$ = empirical constants

   T^1 = temperature

K reaction rate constant can be calculated by equation (4) or reaction (5) D, decimal reduction time can be calculated by equation (6)

To check the values of K, D by using different equations, the organism which has the following 'D' values are taken Microorganism

D = 7.18 min at 115°C

D = 2.27 min at 126.5°C

Δ S* value assumed to be independent of temperature in equation (5)

The calculated values of D values are plotted against T in Fig (8.5)

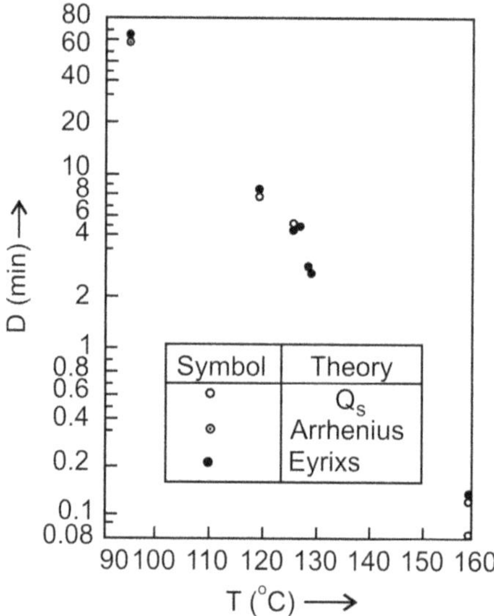

Fig. 8.5 Comparison of calculation of decimal reduction time by, Q_{10}, Arrhenius and Eyring theories.

This plot values show that D values calculated by using K values of Arrhenius theory and Eyring theory are same. While D values calculated by Q_{10} cannot be used over a wide range of temperatures.

The values of activation energy 'E' is in the range from 50–100 k cal / g mole to vegetative cell and spores respectively. For enzymes and vitamins, these E values are in 2–20 kcal / g mole range.

8.2   Design of Sterilizing Equipment

Life span distribution of spores

The spores life span may be defined as the length of time spores remain viable after exposure to given particular temperature.

The specific life span can be attributed to each bacterial spore, whereas distribution of their life span to the suspension of bacterial spores.

The N (number of viable spores) values are generally of the order of 10–100. Also the death rate thus observable is an overall or deterministic value which is observed in plots fig (1 and 2) of N/N_o vs time. While determining the life span of spores, probabilistic approach is used and for media sterilization generally $N \leq 1$.

Let N_o is spore population containing

N_i number and ti = life span at specific temperature

Distribution of life span by introducing probability associated term Pr can be expressed as

$$P_r = \frac{N_o \,|}{\pi N_i \,|} \cdot \frac{1}{K} \qquad \dots\dots(7)$$

Where k = normalization factor

$$N_o = \sum_{i=1}^{\infty} N_i \qquad \dots\dots(8)$$

$$i = \frac{1}{N_0} \sum_{i=1}^{\infty} N_i \, t_i \qquad \dots\dots(9)$$

i = mean value of life span

Since as N_o is generally large it most common life span distribution, P_r (maximum value) may be calculated by applying Lagrange method to equation (7) & (9)

$$N_i = \acute{\alpha}\, \exp(-\beta'\, t_i) \qquad \dots\dots(10)$$

$\acute{\alpha}, \beta'$ are empirical constants

As N_i and t_i are continuous and not discrete then equation (10) may be modify as

$$-\frac{dN}{dt} = \alpha'\exp(-\beta't) \qquad \dots\dots(11)$$

It is known from equation (11) that the most probable pattern is the experimental distribution of life span of individual.

From eqn (8) and (9)

$$N_o = \int_0^{\infty} -\frac{dN}{dt}\,dt \qquad \dots\dots(12)$$

And
$$\bar{t} = \frac{1}{N_o} \int_0^{\infty} -\frac{dN}{dt}\,t\,dt \qquad \dots\dots(13)$$

We have that after substitution eqn (11) in the eqn (12) and (13)

$$\acute{\alpha} = N_o / \bar{t} \qquad \dots\dots(14)$$

$$\beta' = 1/ \bar{t} \qquad \dots\dots(15)$$

eqn (11) (14) & (15), we have

$$-\frac{dN/dt}{N_0} = -\frac{1}{t}\exp(-\frac{1}{t}t) \qquad \dots\dots(16)$$

Upon integration of eqn (16), we have

$$N = N_o \exp\left(-1/\bar{t}\ \ t\right) \qquad \qquad .....(17)$$

The reaction rate constant K min^{-1} in thermal death of spores can be given as the reciprocal of mean life span of single spores for eqn (2) and (17) so

$$K = 1/\bar{t} \qquad \qquad .....(18)$$

Bacterial Spores-Average Life Span

The fraction 'm' of spores has life span $< t$ at a given temperature from eqn (16) and (18) can be as

$$m = \int_0^t K e^{-Kt} dt = 1 - e^{-Kt} \qquad \qquad .....(19)$$

The probability of a specific spore selected randomly from a population, having less than 't' is experiment as

$$P = 1 - e^{-Kt} = 1 - \exp\left(-\frac{1}{t}t\right) \qquad \qquad .....(20)$$

So the probability P(t)

The probability P(t) of the life spores n total is $< t$ at a given temperature, is given as

$$P(t) = (1 - e^{-kt})^n \qquad \qquad .....(21)$$

Now considering specific lump containing n single spore, the life time distribution f(t)n dt between t and t+dt is

$$P(t) = \int_0^t f(t)_n dt \qquad \qquad .....(22)$$

$$F(t)_n = \frac{dP(t)}{dt} \qquad \qquad .....(23)$$

Substituting eqn(21) and (23) , we have

$$F(t)_n = n(1 - e^{-kt})^{n-1} k e^{-kt} \qquad \qquad .....(24)$$

$\bar{t}_{tn}$ = mean value of life span for eqn (24) is therefore given as

$$\bar{t}_{tn} = \int_0^\infty f(t)_n t \, dt$$

$$t_n = \frac{1}{k} \sum_{r=1}^n \frac{1}{r} \qquad \qquad .....(25)$$

And the eqn (25) we have

$$\frac{t_n}{t_1} = \sum_{r'=1}^{n} \frac{1}{r'} \qquad \qquad .....(26)$$

So, for high values where $n \geq 10$ and rhs is given as

$$\frac{\bar{t_n}}{t_1} = \gamma + 2.303 \log n$$

$$= 0.577 + 2.303 \log n \qquad \qquad .....(27)$$

Where γ = Eulers constant

An example taking $K = 1$ min^{-1} for life span distribution calculated is given in Fig. 8.6.

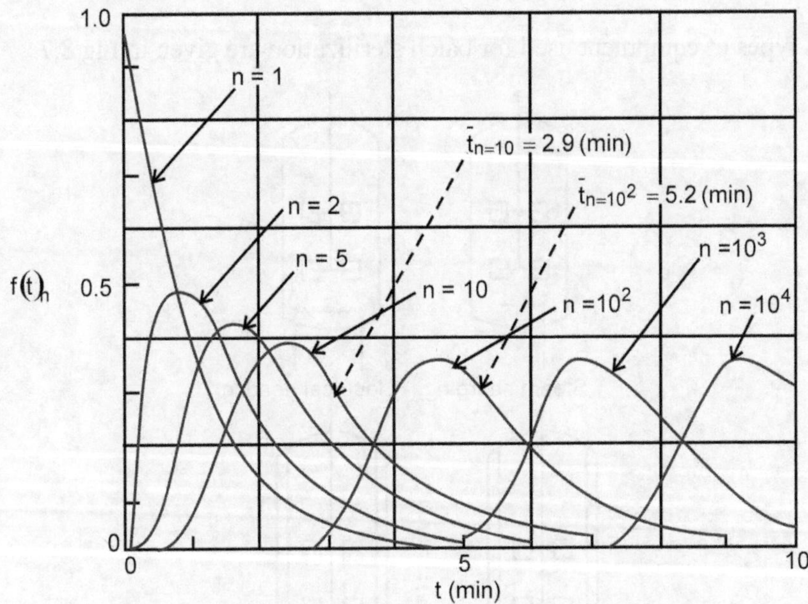

Fig. 8.6 F $(t)_n$ Vs t   ; k = 1 min^{-1}.

It is observed for the fungi the tn for life span of the bacterial lump as the single spore n is added to it.

8.3   Batch Sterilization

In batch sterilization the contents are fed into the sterilizer in one shot at the beginning and steam is injected to carry-out sterilization. One disadvantage, that is expected is, destruction of some nutrients along with the contaminants.

From the equation (1) and (4)

$$\frac{dN}{dt} = -KN = \alpha' e^{-E/RT} N$$

Upon the integration

$$V_{total} = \ln \frac{N_o}{N} = \int_0^t Kdt = \alpha \int_o^t {}'e^{\frac{-E}{RT}} N \qquad\qquad .....(28)$$

Where v^- = design criteria

V^- is called as 'Del factor', Nabla factor or sterilization. The del factor may be defined as the measure of fractional reduction in living organism count over the initial number present, and produced by a heat and time regime.

Different types of equipment used for batch sterilization are given in Fig 8.7

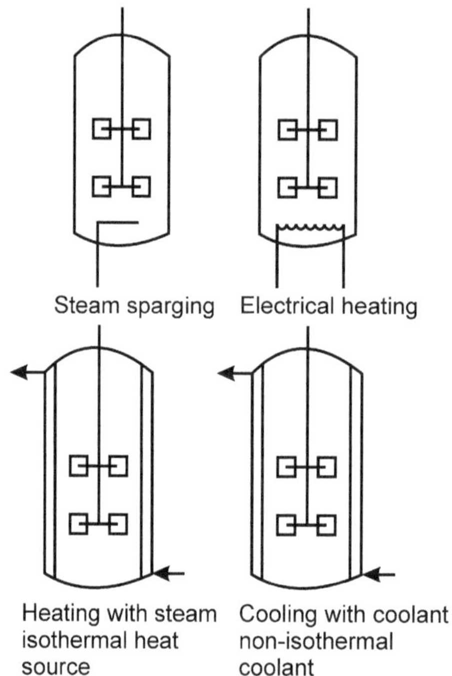

Fig. 8.7 batch sterilization media equipment.

Design equation required for calculating temperature – time profile for each equipment is given in the Table 8.1.

Table 8.1 Temperature and time profile of batch sterilization

Typing of heat transfer	Temperature time profile	á, β,γ
Steam sparging	$T = T_0(1 + \alpha/(1+\gamma t))$ …(29)	$\alpha = hs/(M\,c_p\,T_o)$
Electrical heating	$T = T_0(1 + \alpha t)$ …(30)	$\acute{\alpha}, = q/(Mc_pT_0)$
Heating with steam	$T = T_0(1 + \beta e^{-kt})$ …(31)	$\alpha = U\,A/\,Mc_p$ $\beta = T_0 - T_M/T_M$
Cooling with coolant	$T = T_{co}(1 + \beta e^{-\alpha t})$ …(32)	$\alpha = \left(\dfrac{wc^1_p}{Mc_p}\right)\left\{1 - \exp(\dfrac{-UA}{wc^1_p}\right\}$ $\beta = \dfrac{T_0 - T_{co}}{T_{co}}$

Eqn (28) can be subdivide into

$$\nabla_{total} = \ln\frac{N_0}{N_1} = \nabla_{heating} + \nabla_{holding} + \nabla_{cooling} \qquad\qquad …..(33)$$

$$\nabla_{heating} = \ln\frac{N_0}{N_1} = \int_0^{t_1} K\,dt = \alpha'\int_0^{t_1} e^{-\frac{E}{RT}}\,dt \qquad\qquad …..(34)$$

$$\nabla_{holding} = \ln\frac{N_1}{N_2} = \int_0^{t_2} K\,dt = \alpha'\int_0^{t_2} e^{\frac{E}{RT}}\,dt = \alpha'e^{\frac{-E}{RT}}t_2$$

$$\nabla_{cooling} = \ln\frac{N_2}{N} = \int_0^{t_3} K\,dt = \alpha'\int e^{\frac{E}{RT}}\,dt$$

$$t = t_1 + t_2 + t_3$$

N = sterility level (No of continuous organism after sterilization)

N_o = Continuous after level (No of continuous organism before sterilization)

N_1 = Number of continuous organism after heating period t_1

N_2 = Number of continuous organism after heating period t_2

8.4 Media Sterilization-Continuous Mode

The equipment for continuous sterilization of media is given in Fig. 8.8(a) and (b).

(a)   Injection type sterilization

For carrying out moist heat sterilization the steam will be injected directly into the steam be injected directly into the raw medium and the sterilization temperature will be reached immediately without delay. The medium will be held at the sterilization temperature, for sometime that depends on the length of the pipe in the holding section.

The medium sterilized will be cooled instantly by passing through the expansion value into the vacuum chamber. The temperature – time profile for injection type and plate heat exchange type sterilization is given in the Fig. 8.9(a) and (b) respectively.

Temperature and time profile for (a) injection type sterilization

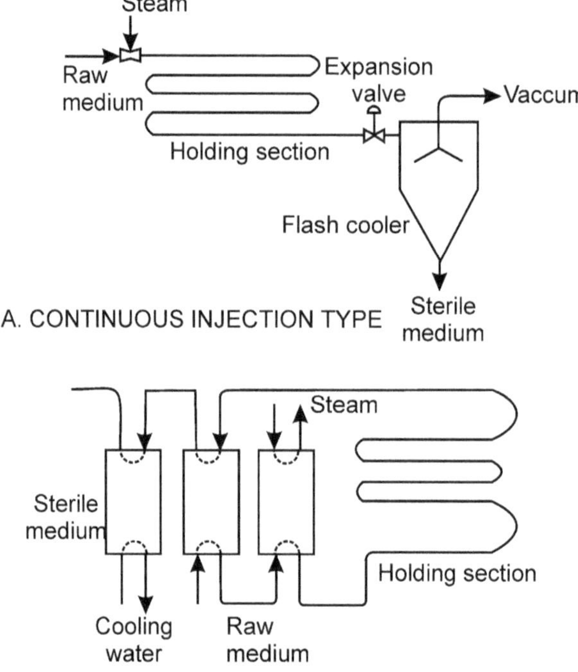

Fig. 8.8 (A) continuous injection type sterilization
(B) continuous plate heat exchange type sterilization

(b)   Plate heat exchange type sterilization

The Fig. 8.9(a) shows that the temperature could be rised to 140°C with out serious damage to medium due to short period of exposure to heat.

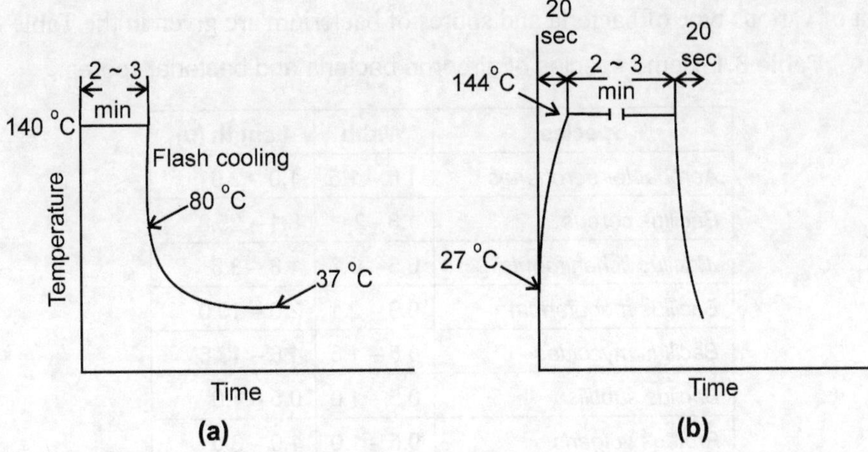

Fig. 8.9 (a) Temperature and time profile for injection type sterilizer,
(b) Plate heat exchanger type sterilizer

Plate heat exchange type sterilization : This type of continuous sterilization is used commonly in fermentation industry. The plate of the heat exchange is heated by the heat source like steam and the hot plates in turn will heat raw medium. Hence the raw medium will be maintained at the elevated temperature for a certain time and then it is cooled in the section of the plate heat exchange.

The time required to heat and cool the raw medium in the plate heat exchange type sterilization is much larger than the steam injection type sterilization. The sterilization time required in the continuous sterilization is much smaller i.e., about 1-2% of the batch sterilization time.

8.5   Sterilization of Air

Sterile air is a process requirement in aerobic fermentation process. Also requirement will be large in case of commercial bio-chemical process industries. The generation of such a large quantity of sterile air is a challenging task for biochemical engineering. Cotton plugs are frequently used in the laboratory experiment whether for test tubes or for shake flasks. Also in pilot plant bioreactors, these plugs do have not any serious contamination problem. Nevertheless commercial filters do not show much advantage.

The filter beds packed with fibers or granular particles do not have 100% collection efficiency to fills the air borne microorganisms. These filters are good enough to prolong the passage time of air borne organism to the next air borne organism. This sort of delay in passage of air borne microorganism may be sufficient to protect the fermentation, particularly during the growth phase of microorganism.

The list of various type of bacteria and spores of bacterium are given in the Table 8.1

Table 8.1 Some species of airborne bacteria and bacterial spores.

Species	Width	Length (µ)
Aerobacter aerogenes	1.0 – 1.5	1.0 – 2.5
Bacillus cereus	1.3 - 2	8.1 – 25.8
Bacilus licheniformis	0.5 – 0.7	1.8 – 3.3
Bacillus megaterium	0.9 – 2.1	2.0 – 10.0
Bacillus mycoides	0.6 – 1.6	1.6 – 13.6
Bacilus subtilis	0.5 – 1.0	0.5 – 1.0
Proteus vulgaris	0.5 – 1.0	1.0 – 3.0
(Spores)		
Bacillus megaterium	0.6 – 1.2	0.9 – 1.7
Bacillus substilis	0.5 – 1.0	0.9 – 1.8

The air borne microorganism vary from several million microns to several hundred million microns in size. Interestingly the small air borne microorganisms are adsorbed by the dust particles in the air and they can easily be removed along with larger dust particles. The sizes of the particles which vary from $0.5 - 1.0\mu$ are either destroyed or collected during air sterilization. The medium size of the dust particles of about $0.6\ \mu$ are normally found in both indoor and outdoor air.

The number of organisms per m^3 of air n was calculated after incubation for $2 - 3$ days Fig. (8.10).

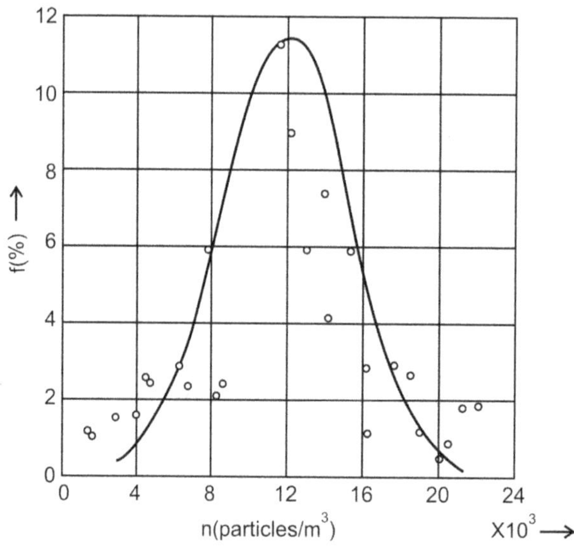

Fig. 8.10 A typical plot frequency distribution of organisms in air.

If N is the total number of microorganisms that are found in large volume of air, the probability P (n) with an unit volume m^3, selected from the bulk air, will contain n microbes and can be expressed as

$$P(n) = (\frac{N}{n})P^n(1-P)^{N-n} \qquad .....(37)$$

P = probability with which each microorganism out of N is found in the unit volume air which is selected

Methods of Air Sterilization

Air sterilization can be achieved by adopting various means like heat, U-V rays, germicidal sprays and mechanical filtration.

Heat

Generally the bacterial spores are resistant to heat but they will be destroyed if high temperature is applied. Some studies reported that at temperature above 200°C for short time, bacterial spores were destroyed. Also the heat generated by air compressor which is about 200 °C could also destroy the spores of bacteria.

Ultra Violet Rays

Ultra violet rays having wavelength from 2,265 – 3287 A° were found to be efficient in killing microorganisms. Mostly these U-V rays are used in sterile room of biotechnology based industries and operation theatres in hospitals.

Germicidal Sprays

Germicides like ethylene oxide, phenol or salts of heavy metals may be sprayed in air conditioning equipment to kill the air borne bacteria. The efficiency of such methods may not be 100%.

Mechanical Filtration

Various types of mechanical filters have been used like fibrous air filters. Initially cotton fibres were used for air sterilization and they are replaced by glass fibres. Also it is observed that the polyvinyl alcohol filter (PVA) are resembling glass fibers filters in their performance. Polyvinyl alcohol type filters are replacing conventional type of filters.

The air sterilization is achieved by the combination of heat and mechanical filtration in process industries.

Fibrous Type Filters

The different mechanical adopted by fibrous type air filters for removal of microorganism are

   (a)   inertial impaction
   (b)   interception
   (c)   diffusion
   (d)   settling by gravitational force
   (e)   electrostatic force

Now consider the filtration of contaminated air through fibrous filter.

Air borne organisms may escape the filter when it does not hit fibrous material as the particle size is smaller than the gap in between the fibers. Also when the particle or organism, when attached to filter, clean air is passed through the filter. So uniform concentration of particles may be assumed throughout the filter.

This may be explained by

$$\frac{dN}{dt} = -KN \qquad \qquad .....(38)$$

Where N = number of particles at a depth X of filter

$\qquad K$ = constant

Upon integration of eqn(38) we have

$$\frac{N}{N_0} = e^{-kx} \qquad \qquad .....(39)$$

N_0 = Number of particle at the entrance of the filter bed at $X = 0$

Equating (39) may be written as

$$\ln \frac{N}{N_0} = -KX \qquad \qquad .....(40)$$

The efficiency of the filter may be defined as the ratio of the number of particles moved to the initial number of particles,

Efficiency may be expressed as

$$E_f \frac{N_0}{N_0} - \frac{N}{N_0} = 1 - \frac{N}{N_0} \qquad \qquad .....(41)$$

$$E_f = 1 - e^{-KX} \qquad \qquad .....(42)$$

X_{90} may be defined as the depth of filter required to remove 90% of solids entering the fibrous filter ie $E_f = 0.9$

From eqn (40) and (42) we have

$$X_{90} = 2.303/(k) \qquad \qquad .....(43)$$

K is function of filter medium and air velocity

So if the air velocity is high, K will initially increase, mainly due to high air velocities.

The filter bed has to be designed so that the particles will be less in outgoing sterile air.

Sterilization of Bioreactors

When sterile medium is separately introduced into a bioreactor, the bioreactor has to be sterilized prior to sterilization of medium. Nutrients and microorganism may stick to pits or flange joints inside the bioreactor. Steam sterilization is adopted to sterilize the bioreactor i.e., steam at 15 psig is passed and maintained for 20 mintuties. Bioreactor has to be flushed with sterile air under positive pressure after the steam sterilization.

Normally, the joints inside the bioreactor, are kept minimum to ensure sterile conditions.

References

1. Aiba, S. Humphrey A.E, and Millis N.F (1973) Biochemical engineering Academic press, New York

2. Banks, G.T (1979) Scale up of fermentation processes Topics in Enzymes and fermentation Biotechnology Wise man A (ed) vol. 3, Ellis Horwood, Chichestr.

3. Humphrey, A.E., and Gaden, E.L. (1955), "Air sterilization by fibrous media", IEC.

4. Richards, J.W. (1968) Introduction to Industrial sterilization Academic press, London.

5. Stanburry, P.F and Whittaker, A. (1993), Principles of Fermentation Technology, Pergamon Press, Oxford.

Review Questions

1. Why sterilization has to be carried out in a bioprocess?
2. Explain thermal death of microorganisms.
3. Write about temperature effect on death rate.
4. Explain batch sterilization.
5. Describe continuous media sterilization.
6. Explain sterilization of air.

CHAPTER 9

Industrial Fermentation

Industrial fermentations have been practiced since ages. Even through the fermentation technology was not developed, the production of alcoholic beverages was carried out. Industrial fermentation can be broadly divided into anaerobic fermentation and aerobic fermentations. Anaerobic fermentation include, ethanol production, lactic acid production, and acetone ethanol fermentation, while the aerobic fermentation include, citric acid production, bakers yeast production, penicillin production, HFCS (High fructose Corn Syrup) and production on single cell protein. In all these process the products have been produced by using microorganisms and various suitable substrates, like glucose, sucrose and starch etc. The conditions required for the fermentation varies from one process to another. Since technology is changed over a period of time, high value and low volume products have been produced by using recombinant DNA technology. The scope of this book is confined to contain aerobic and anaerobic fermentations only. Products of commercial importance are described here.

9.1 Production of Penicillin

The antibiotic penicillin was discovered by Alexander Fleming in 1928. The organism was found to be *Penicillin notatum*. This culture stopped the growth of *Staphylococcus aureus* on agar against gram positive bacteria but it does not act on gram negative bacteria. Since it attacks the synthesis of cell wall as observed, it is active against the cells that are growing. Various types of pencillins are available in market. The structure of penicillin 'S' and penicillin 'V' are given in the Fig. 9.1.

Fig. 9.1 chemical structure of sodium penicillin G.

Fig. 9.2 chemical structure of penicillin 'V'.

Since sodium penicillin- G is degraded in acidic conditions, it is administered parentally. Also (a) penicillin-'V' is acid stable, it is administered orally. Various forms of penicillin can be produced by using different genes of penicillin organism. Normally these organisms may be stored in lyophilizers and the vegetative cells may be suspended in glycerol and stored at -70^0C.

Penicillin production process is descried in the Fig.9.3. Variety of carbohydrates like, sucrose, glucose, hydrolyzed starch, lactose and molasses are used as carbon sources by *P. chrysogenum*, while corn oil may be used as energy source. Nitrogen sources are like CSL (corn steep liquor) are important as they contribute to the high yields of penicillin. Precursors of penicillin side chain like sodium phenyl acetate and sodium phenoxy acetate per gram of penicillin should be added to fermentation medium to produce penicillin 'G' and penicillin 'V' respectively.

Penicillin is secondary metabolite and during initial phase, cell growth is expected.

The process of penicillin production is described as given below.

(a)   Fermentation of Penicillin

1.   First 40 hours of fermentation exponential cell growth is expected with t_d = 6 hrs.

2.   Nutrients glucose and corn steep liquor and precursors like sodium phenyl acetate are added when cell density is increased to maximum production of penicillin.

3.   Culture preparation starts by using lyophilized culture and agar slant cultures.

4.   Vegetative cells are developed in shake flaks to transfer then to seed fermentation of volumes 10-1000 *l*.

5.   The contents of seed fermentation after maintenance are transferred to production fermentation of volumes 200-500m^3.

6.   The operating conditions of production fermentor are, agitation 100-300 rpm and temperature at 25-28 ^{0}C is maintained by cooling at, D.O (dissolved oxygen) marked > 2 mg/l and pH at 6.5.

7.   Fermentation is stopped when oxygen uptake rate of culture exceeds oxygen transfer rate of the reactor.

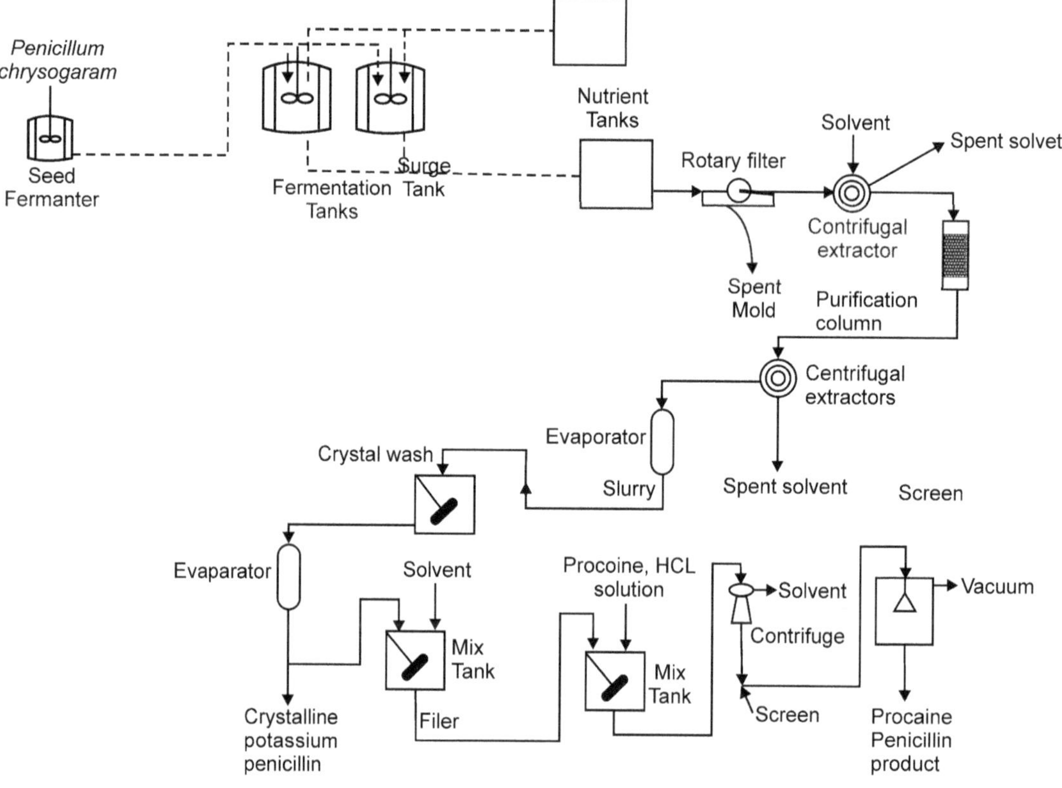

Fig. 9.3 Flow diagram of penicillin production.

(b) Recovery of Penicillin

The recovery process includes filtration, extraction, adsorption, crystallization and drying.

1. Filtration is carried out by using rotary vacuum drum filters. Mycelia deposited on the filter are washed off. So the penicillin rich filtered is cooled to 2–4^{0}C, to avoid enzymatic or chemical degradation of penicillin.

2. Solvent extraction is carried out at pH of 2.5 – 3.0 by using solvents like amyl acetate or butyl acetate.

3. Centrifugation is carried out by using continuous counter current multistage extractors.

4. The distribution coefficient of penicillin 'G' or 'V' between organic and aqueous phases depends strongly on the pH of the medium. So the distribution coefficient is 20 at pH of the medium. So the distribution coefficient is 20 at pH-3.

5. To avoid penicillin degradation at pH-3, acidic solvent extraction, the temperature is maintained about 2 – 4^{0}C while filtration temperature is kept small i.e., 1-2 min.

6. 99% Penicillin is recovered by using two extractors.

7. Impurities are removed by using carbon adsorption from penicillin rich extract.

8. Crystallization is done from solvent or aqueous phase. 'Penicillin' concentration, Na, K, pH and temperature is to be adjusted for crystallization. Na, or K are added in excess amounts. World wide, approximately 26000 tons of penicillin 'G' and 10,000 tons of penicillin 'v' are produced per annum.

9.2   Production of Citric Acid

Citric is produced by *Aspergillus niger* is used for the production of citric acid. The uses of citric acid include beverage, confectionaries, pharmaceuticals and industry. Normally the carbohydrate sources used for citric acid production are beet root or cane molasses. Citric acid is a primary metabolite, produced under nitrogen and phosphate limitations and after growth has ceased. The production process is described briefly as given below.

1. The specific condition required per citric acid production are,

   (a) Restricted growth

   (b) Medium deficient in one or more essential elements

   (c) High sugar concentration

   (d) High dissolved oxygen concentration

   (e) pH < 2

   (f) Absence of trace elements.

   (g) Oxalic acid gluconic acids are produced if the pH > 2.

2. The concentration of metals is reduced by addition of potassium ferro cyanide and citric acid production is promoted by potassium ferro cyanide, the growth inhibitor.

3. High Dissolved oxygen concentrations are to be maintained during process and acid production is decreased at low oxygen levels.

Surface Fermentation

1. Power requirements are low.
2. The liquid depth is of 5 – 20 cm.
3. Fermentation conditions : beet molasses, sugar concentration of 150g/l ; pH = 6 is taken in trays, Temperature = 30^0C
4. Alkali ferro cyanide is added to the medium.
5. Trays are emptied after 7 – 10 days.
6. Mycelia are removed and medium is taken for recovery.

Submerged Fermentation

1. Deep stainless steel reactors are used of volumes $100m^3$.
2. Fermentors are mechanically agitated.
3. Aeration rate is about $0.1 - 0.4$ vvm, agitation is $50 - 100$ rpm to avoid shear to molds.
4. Nutrients are added to fermentors.
5. pH adjusted 2.5 to 3.0.
6. Spores of *A.niger* are allowed to germinate in inoculum medium and then transferred to main fermentors or bioreactors.
7. Oxygen enriched air may be introduced.
8. Batch operation results in production of $0.5 - 1$ kg/ m^3.hr .
9. After 4 - 5 days, the biomass is separated by filtration; the fermentors are emptied. The liquid is transferred to recovery section.

Recovery

1. Precipitation is done by adding calcium hydroxide (lime) to heated fermentation broth and calcium citrate tetrahydrate is obtained.
2. Precipitate is then washed and treated with dilute sulphur acid, yields solution containing citric acid and $CaSO_4$ (gypsum) precipitate.
3. After bleaching and crystallization either anhydrous monohydrate, citric acid is obtained.
4. The demand for citric acid is about 400,000 tons per annum.

9.3   Production of Bakers Yeast

Long back the leavening of the bread is achieved by inoculating fresh dough with the left over dough of the previous batch of bread making. This type of inoculation resulted in variations in carbon dioxide evolution due to sugar fermentation. Subsequently after sometime residual yeast from distilleries and breweries, was added to the dough but this resulted in bitter taste, due to presence of hops. This was rectified by growing yeast separately for this purpose and brewing medium containing malt, was used without hops. However the present day practice includes molasses, the cheap raw material as medium and sufficient aeration is given to produce more yeast and to avoid production ethanol.

Now a days specially selected organism *Saccharomyces cerviseae* are used for bakers yeast production.

Bakers yeast is produced by aerobic fermentation

1. The inexpensive raw material and carbohydrate sources is molasses and other carbohydrate sources are glucose, sucrose, fructose, or hydrolyzed starch. Nitrogen sources are urea, ammonium salts and aqueous ammonia.

2. The temperature is controlled at 30^0C and pH is maintained at $6 - 7$. The dissolved oxygen concentrations are maintained above 2 mg/L.

3. The broth is sent for centrifugation to separate yeast after 20 to 30 hrs of culture.

4. Cells are washed to remove inert solids light colored cream is produced due to centrifugal separation.

5. Cream is stored in agitated tanks at $2 - 4\ ^0C$ and part of this cream is used as seeding material for other batches of fermentation.

6. Bakers yeast may be sold in the form of cream, or in compressed form (30% solids) or dried form (95% yeast solids).

7. The cream may be concentrated either in filter press or in rotary vacuum filters. Emulsifiers are added to the filter cake and then extruded to yeast cakes.

8. Active dry yeast is produced by exposing yeast cake to hot air stream. The temperature of the drying is kept at 45^0C, since above 50^0C these vegetative cells are killed. The final moisture content of yeast would be $5 - 10$ % and drying time varies from 20 min to several hours.

9.4   Production of Ethanol

The demand for industrial alcohol is increasing due to its application as motor fuel i.e., by mixing with petrol and mostly as industrial solvents. About four million tones of alcohol is produced annually all over the world. Out of the 4 million tones, 80% is produced by using fermentation process. (Fig 9.4)

1. *Saccharomyces cerviseae* is the microorganism used for the production of ethyl alcohol. *E. coli* may be genetically modified to produce 43% (vol/vol) concentration of ethanol.

2. In India cane sugar molasses is used as a raw material for the production of ethyl alcohol. In Brazil also, cane sugar molasses, is used as a raw material for the production of ethanol, while corn is used as a raw material in United States.

3. The overall reaction for ethanol production may be written as (glucose conversion to ethanol)

$$C_6H_{12}O_6 \quad \text{---------------}> \quad 2C_2H_5OH + CO_2$$

4. The temperature maintained during fermentation is $30 - 35^0C$ and the pH is maintained at $4 - 6$ and redox potential is kept below $- 100$ mv by deploying reducing agent such as Na_2S.

5. The feed solution to the fermentor or bioreactor should include nitrogen sources, phosphorous sources minerals along with trace elements.

6. A Batch fermentor is carried out for 30 – 40 hours and productivity is 6g/liter. hr.

7. In continuous fermentation about 95% sugar will be converted to ethanol and residence time will be about 10 hours. Continuous fermentation with cell recycle will have a productivity of 30 g/*l*. for feed glucose of 100 g/l. Continuous fermentation with recycle is shown in Fig. 9.4.

8. Ethanol will be separated from the fermentor by applying low temperature vacuum distillation, adsorption or membrane separations.

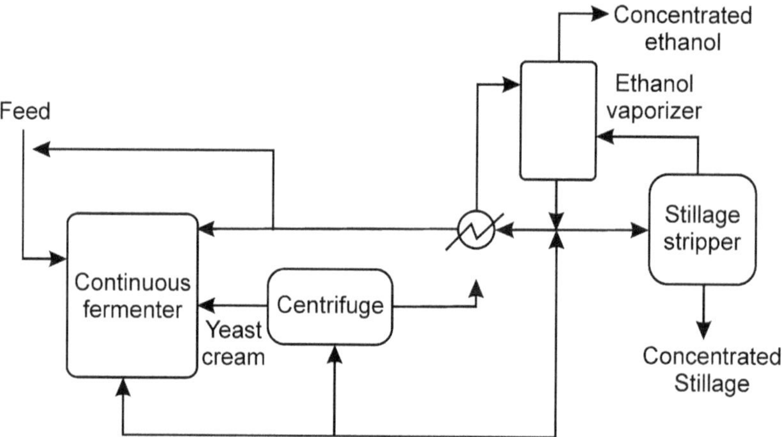

Fig. 9.4 Ethanol production by continuous fermentation.

9.5   Production of Single Cell Protein

Single cell proteins are nothing but the dried cells of the microorganisms. Of course, microorganisms mostly contain proteins. The single cell proteins (SCP) are used as animal or human foods and the examples are yeast, bacteria, molds, actinomycetes and algae. The single cell protein are used for human consumption include algae (spirulina) taken from ponds and lakes, yeast in leavened bread, and lactic acid bacteria in fermented dairy products in sugars, cheese etc.

Most of the SCP is used as in animal food supplement and human consumption is at low level only. Most of the nations in the world are involved in SCP production.

SCP production by using *Methylophilus methylotrophus* bacteria is given in the Fig. 9.5.

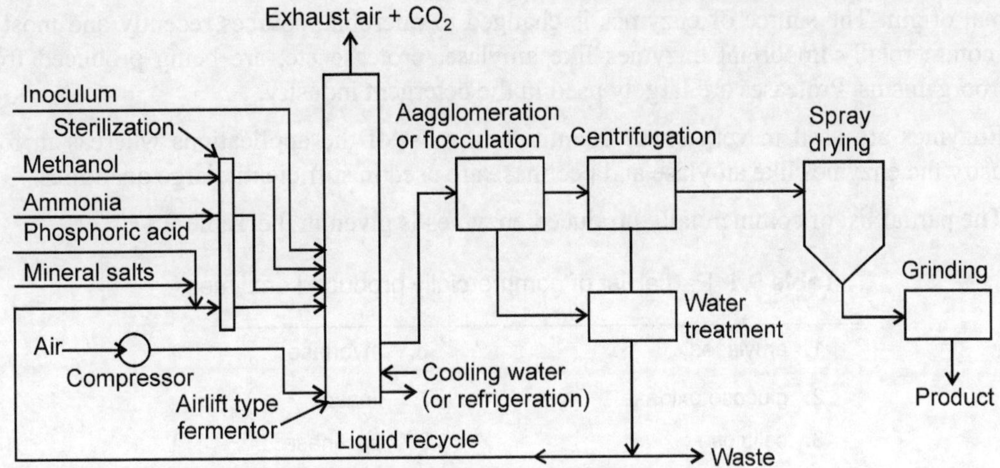

Fig. 9.5 Flow diagram for SCP protein.

$$1.72CH_3 + 0.2\ NH_3 + 1.51\ O_2 \text{-----------}>\quad 1.0CH_{1.68}O_{0.36}N_{0.22} + 0.72\ CO_2 + 2.94\ H_2O$$

The microorganisms grow rapidly with a specific growth rate about 0.5 hr^{-1} and the broth becomes dense by the end of fermentation. The dense fermentation broth is then fed to centrifuge. The growth yield is above 0.50 g cell / g substrate. SCP production will be increased by using improved strains of *S.cervaisiae* and also by implementing automatic control systems. Torula yeasts are commonly used as food grade SCP which impart flavor and texture.

9.6 Vaccines

Vaccines promote the host in producing antibodies and induce immunity against infectious disease. Now-a-days, microorganism are used for producing vaccines of regular use like Hepatitis vaccines which is used against Jaundice, a liver malfunctioning disease. Vaccines are being produced against rabbis, anthrax etc.

For example *Bacillus anthraces,* the causative agent of anthrax, possesses two main virulence factors, a poly – D glutamic capsule and a tripartite protein toxin. The production of immunogenic   proteins such as the protective antigen (PA) of *Bacillus anthraces* using recombinant DNA technology is gaining commercial importance.

Bacillus subtilis is a harmless bacteria existing in environment. B.E. Ivins et. al., proposed the possibility of using a genetically transformed *B.subtilis* to production PA, without other undesirable components of the anthrax toxin. The detail production process of vaccines is out side the scope of the Book.

9.7 Enzymes

Enzymes basically contain proteins. These enzymes were known since long time i.e., as early as 800 B C, mostly for cheese production. Enzymes were produced either from plant or animal origin. The source of enzymes is changed to microbial sources recently and most of the commercially important enzymes like amylase, protease etc, are being produced from microorganisms. Proteases are largely used in the detergent industry.

Enzymes are used in very small quantities in most of the applications whereas in food industry the enzymes like amylase and pectinase are used in sufficiently large quantities.

The partial list of commercially produced enzymes is given in the Table 9.1.

Table 9.1 Partial list of commercially produced enzymes.

1. amylases	6. invertase
2. glucose oxidase	7. lipase
3. cellulose	8. pectinase
4. catalase	9. protease
5. Hemi cellulase	10. pencilllinase

The enzymes produced commercially are used in various sectors like industry, used for analytical purpose and in the medicine. The application of enzymes are gives in Table 9.2.

Table 9.2 Applications of enzymes

S.NO	ENZYMES	APPLICATION	SAILETENT FEACURES
1.	Catalase, Amylases, Proteases, Isomerases, etc.	used in various industries, textiles detergent etc,	Microbial origin mostly extracelluar
2.	Cholesterol Oxidase Hexokinase, Alcohol dehydrogenase	Analytical purpose	Microbial animal and plant source mostly intracellular
3.	lipase, protease Strepto kinase	used in the field medicine	Microbial animal and plant sources mostly Intracellular.

Protein enzymes are largely used in the detergent industry, world-wide, apart from dairy products fruit juices, etc. The applications of enzymes are increasing due to availability of pure enzymes and in large quantities.

9.8   High Fructose Corn Syrup

The application of high fructose corn syrup is increasing due to its wide applications in sweet foods, beverages and in desserts. Since in 1940 the corn syrups have been produced enzymatic ally by hydrolyzing corn starch. The conversion of glucose to fructose was commercialized after this commercially produced. HFCS was about 15% fructose concentration initially and later on due to improvements in process, enhanced to 42% and 55% fructose concentration. HFCS is divided in to three major categories basically depending on its fructose concentration i.e., 42% - 55% and 90%. The uses of these three types are given in the Table (9.3).

Table 9.3 Applications of HFCS

S.NO	HFCS having fructose Concentration	Applications
1.	42%	food products using liquid sweeteners.
2.	55%	soft drinks
3.	90%	jams and jellies

The production of high fructose corn syrup is given in the Fig. 9.6.

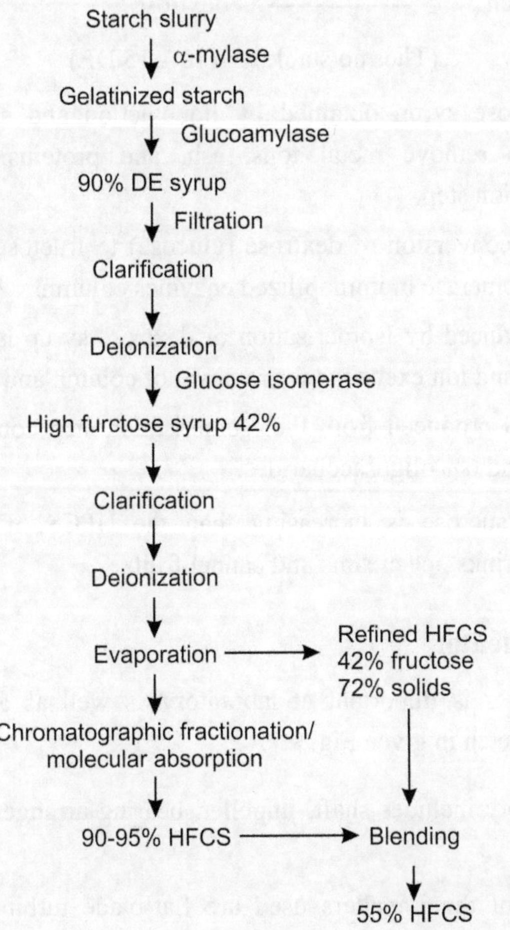

Fig. 9.6 Schematic process diagram of HFCS production.

The five major operation of the process are given below:

1. Enzymatic hydrolysis of corn starch to produce dextrose

2. Treatment of dextrose syrup, primary physical and chemical

3. Dextrose to 42% fructose – Isomerisation

4. Fructose corn syrup – secondary refining

5. Conversation of HFCS 42% fructose to 55% fructose. The process steps of HFCS production is described below :

   (a) Corn starch is subjected to temperature of about 65^0C and is gelatinized to produce dextrose.

$$(\alpha \text{ - amylase})$$

Corn starch   -----------------> dextrose

$$(\text{Thermo stable})   (10 - 15 \text{ DE})$$

   (b) The dextrose syrup obtained by liquefaction and saccharification is further refined to remove metal tons, ash, and proteins. These interfere during Isomerisation step.

   (c) Next, the conversion of dextrose (glucose) to fructose is achieved by enzymes glucose isomerase in immobilized enzymes column.

   (d) HFCS produced by Isomerisation of dextrose syrup is for the refined by colon treatment and ion exchange for removal of column and ions respectively.

   (e) The HFCS produced from the Isomerisation step contains 42% fructose, 52% dextrose and 6% oligosaccharides.

Since the cost of sucrose is increasing than the HFCS, it is replacing sucrose in applications like, soft drinks, ice creams and cannel fruits.

9.9   Design of Fermentor

The design of fermentor is important at laboratory as well as at commercial level. The laboratory fermentor sketch in given Fig. 9.7.

The agitator assembly includes shaft, impeller, bearing arrangements, motor and aseptic seals.

The various types of the impellers used are flat-blade turbine, marine propeller, and V shaped design etc.

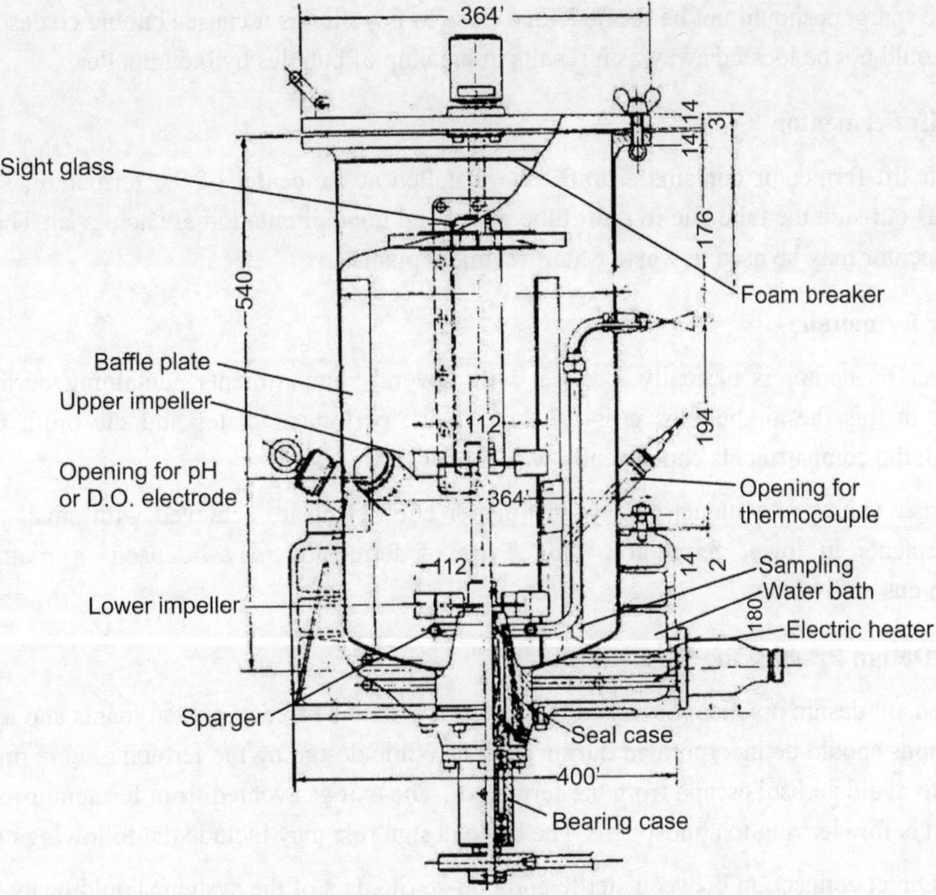

Fig. 9.7 A diagram of fermentor with design details.

Normally two types of spargers are used in fermentor industry i.e.,

1. single hole sparger containing single hole i.e., a pipe located below impeller
2. ring sparger with holes located in diameter are drilled tube periphery at the bottom. The area of holes will be equal cross section area of tube. The superficial velocity vs. and aeration rate concepts are used to circumvent foaming and flooding problems. The examples are given in table (9.4).

Table 9.4 Impeller and superficial velocities

S.No.	Impeller	Superficial velocity	Resulting
1.	Marine propeller	> 21m/h	flooding
2.	Paddle impeller		
3.	Flat blade turbine	>120 m/h	flooding

The spargers should not be located very close to impeller as it causes bubble coalescences and should not be located away as it results in breakup of bubbles by the impeller.

Air Lift Fermentor

The air lift fermentor contains a draft tube installed at the center of the fermentor, so that bubbles out-side the tube due to draft tube effect and good circulation are achieved. The type of fermentor may be used in waste water treatment plants.

Tower fermentor

A tower fermentor is basically a vessel with several compartments containing perforated plates. In this the air bubbles pass–up through the perforated plates and the broth passes through the compartments concurrently with the air flow.

Larger values of volumetric oxygen transfer coefficient are achieved with small power requirements in tower fermentor. This Type of fermentor may be used in multistage continuous cultivation.

Good Design Practice for Fermentor

The general design practice for any fermentor is, it should be free of contaminants and aseptic conditions should be incorporated during design. While designing the fermentor care must be taken to avoid aerosol escape from the fermentor. The exit gas vented from fermentor may be filtered before leaving to atmosphere. The basic design rule may include the following norms.

1. Direct connection between sterile and non-sterile pars of the system should be avoided; since bacteria may grow in inter-connecting values and joints.
2. Flanged connections should be minimized. Since flanged connections move due to equipment vibration and thermal expansion.
3. Welded connections used should be properly polished to avoid contamination by solid medium in the reservoirs.
4. Dead spaces, crevices etc; should be avoided to avert solids accumulation leading to contamination.
5. The parts of the fermentor should be sterilized independently.
6. The connections to the fermentor like sampling ports should live steam connections.
7. Early sterilizable valves like, ball values diaphragm values and globe values should be used.
8. Positive pressure to be maintained in the fermentor so that leakage always out.

Fig. 9.8 Connections for a typical fermentor.

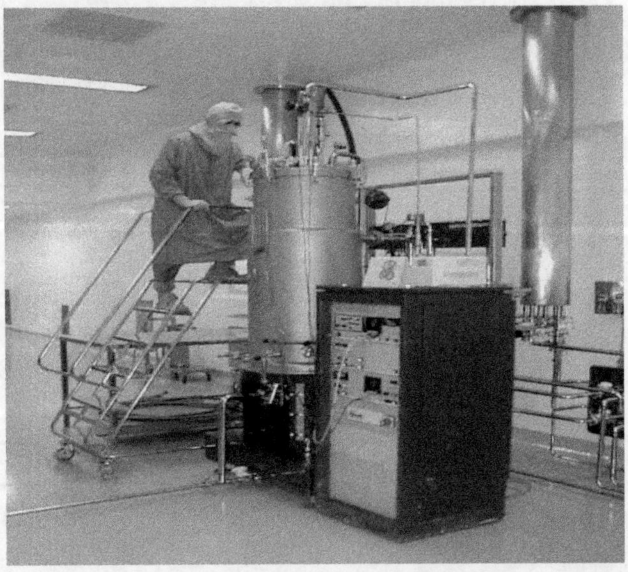

Fig. 9.9 Computer controlled fermentor.

Material of Construction

Mostly either glass or stainless steel has to be used for fermentors. While glass may be used to vessels of 20 – 30 l size, stainless steel may be used in larger size vessels.

Bearing Assemblies

Bearing assembling two types may be used

1. out board (out side vessel) and
2. in board (inside vessel)

Examples external bearing assumptions is given (Fig. 9.10).

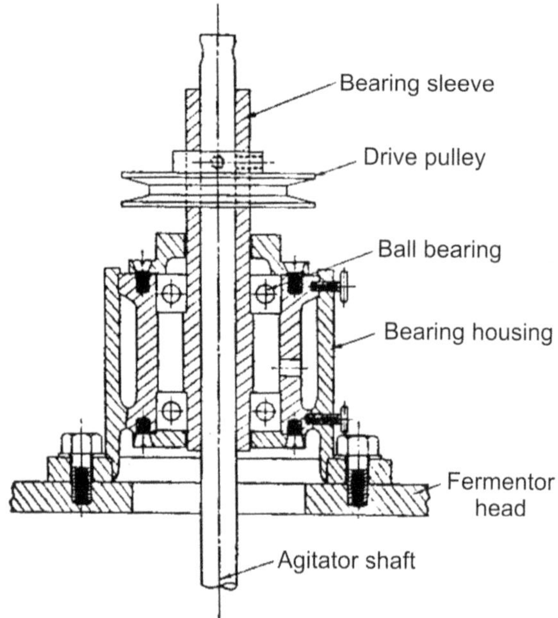

Fig. 9.10 Bearing assembly for a fermentor.

Motor Drives

Power input to a fermentor may be connected as given in the Table 9.5.

Table 9.5 Power input to a fermentor

Size	Power in put (w/*l*)
Laboratory	8 – 10
Pilot plant	3 – 5
Plant	1 – 3 (equivalent to 1- 4 HP/ m^3)

Small electric motor with silicon rectifiers for speed and power control is good for small size fermentor. Often motors controlled by rheostats should be avoided due to gross variations.

Variable speed are provided by using pulleys for intermediate drivers up to 10 HP motors. 2 speed drivers used for large drive > 100HP since variable speed drivers are costly.

Aseptic Seals for Large Fermentors

A jar fermentor with an aseptic seal is given in Fig. 9.11, generally the seals are lubricated before sterilization with a silicone.

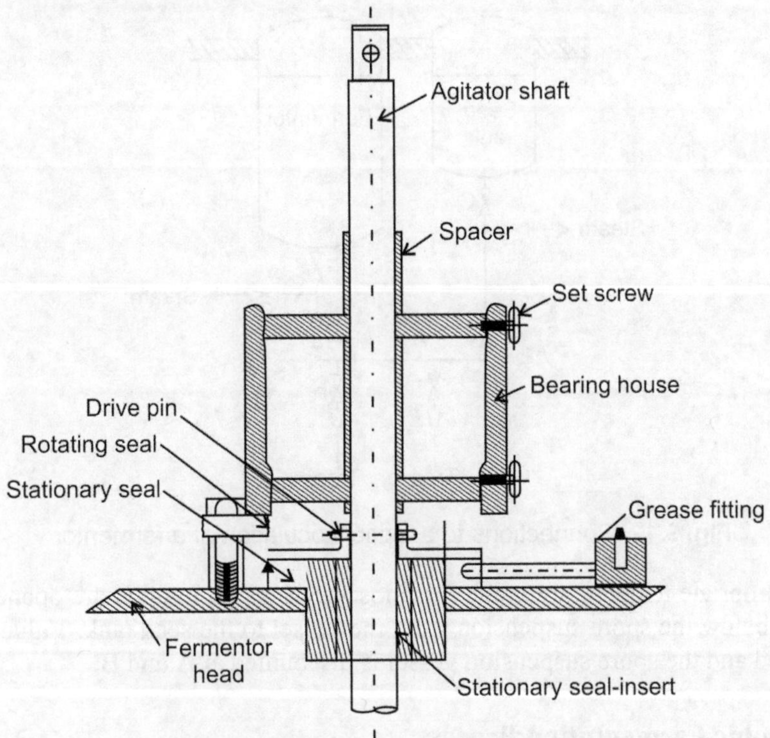

Fig. 9.11 Fermentor with aseptic seal

Grease is used which does not become fluid during sterilization at 120^{0}C for min. The details of other seals are not within the scope of this book.

Aseptic Operation of Fermentor

The pipe lines connected to the fermentation process, for flow of sterile air, medium and other materials should be sterilized with steam at 120°C for 20 to 30 min. Also the condensate of steam should not be present after sterilizations.

Inoculation - Aseptic Mode

For Aseptic transfer of spore suspension to feed tank, connection are given in Figure (9.12).

The piping and vessel for spore suspensions are first sterilized and cooled before spores introduced to the seed. The system is connected at both A and B with a pipeline connected to seed vessel. The connection at A and B are first slackened so that steam bleeds from A and B when steam flows through open valves, E, F and G, with values D, H, I and C closed (Figure 9.12). After sterilization with steam at 120 °C for about 20 min, values E and G are closed and the connection at A and B are tightened. Value D is than opened and the line is cooled to the desired level under positive pressure with sterile air.

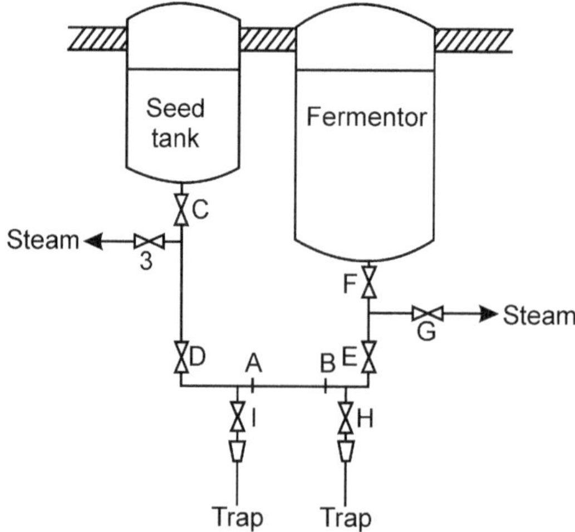

Fig. 9.12 Connections to aseptic inoculation of a fermentor.

At the appropriate temperature value F is closed, values H, I and C are opened and sterile air is used to below the spore suspension from the vessel to the seed tank. Values, C, H and I are then closed and the spore suspension vessel is discounted at A and B.

9.10 Anaerobic Fermentation Process

The example for anaerobic fermentation is alcohol fermentation. Mostly anaerobic organisms are strict anaerobes and they do not require oxygen for their survival. Yeast (*Saccharomyes cerviseae*) is used for the anaerobic fermentation of alcohol. Carbon dioxide is evolved during the alcohol fermentation and it is used for the mixing of reactor contents. So most of the batch fermentor carried out in distillers do not have agitators.

Another example for anaerobic digester is digestion f sludge generated o was water treatment plants. The digested sludge can be used as manure. Also the biogas generated due to anaerobic digestion i.e., mainly containing methane gas and carbon dioxide, can be used for street lighting or power generating electric power.

9.11 Aerobic Fermentation

Most of the fermentations apart from ethanol and acetone butanol fermentations fall under this category of aerobic fermentation. Aerobic microorganisms need oxygen for their survival. For example citric acid fermentation, antibiotic fermentations etc; need dissolved oxygen. Bioreactors have been properly designed to ensure the dispersion of oxygen into the system. Particularly during fermentations, where mycelia are formed or where viscosity of the broth becomes thick during fermentation, a proper air sparging system is essential. Fluidized bed bioreactors are also being tried in addition to conventional CSTR type of bioreactors for carrying out aerobic fermentations.

In addition to the regular aerobic fermentation used for producing microbial products, waste water systems also use oxygen in the biological treatment process. Activated sludge process designed for imparting oxygen either by mechanical surface aerators or by diffused aeration by air compressors. The dissolved oxygen levels are also to be maintained during activated sludge process.

Solid State Fermentation (SSF)

In slid state fermentation the substrate is in solid state and microorganism grows on the surface of the medium and substrate, which is in solid state. Solid state fermentation processes used largely in commercial scale for the production of enzymes. It is also very interesting to note that citric acid is still produced by using solid state fermentation in Maharashtra, India. Of course, due to proprietary in nature, the technology cannot be easily duplicated.

The advantages of solid state fermentation are given the Table 9.6.

Table 9.6 Industrial and economic advantages of SSF

S. NO.	Advantages
1.	low capital investment and recurring expenditure.
2.	low energy requirements for fermentation process due to absence of agitation
3.	low water utilization and hence negligible waste water
4.	No foam formation due to absence of excess water
5.	Simple fermentation media
6.	High reproducibility or results
7.	Less fermentation space, less complex plant and Machinery.
8.	Absence of rigorous control techniques
9.	Any level of scale of operation
10.	No elaborate aeration requirements
11.	Ease in controlling bacterial contamination
12.	Facility of using wet and dry fermented solids directly
13.	Ease in induction and suppression of spores
14.	Lower costs if down stream processing.

Submerged Fermentation (SMF)

The substrate, medium are in liquid state in submerged fermentation. Since the reactor contents are in liquid medium, heat and mass transfer efficient, calling up of the submerged fermentation process is easy.

Some of the salient features of SSF and SMF are given in Table 9.7.

Table 9.7 Salient features of SSF and SMF

Salient features	SSF	SMF
1. condition of microorganism And substrate	Static	Dynamic
2. Status of substrate	Crude	Refined
3. Availability of water	Limited	high
4. Oxygen contact	direct	dissolved
5. Requirement of Fermentation medium	Small	Large
6. Energy requirements	low	high
7. Temperature and concentration Gradients	steep	smooth
8. Chances of bacterial contamination	low	high
9. Pollution problems	low	high

Applications

1. Citric acid can be produce by SSF and SMF. In Maharashtra India one company is producing citric acid by SSF and in Gujarat, citric acid is produced by submerged fermentation.

2. Soya based foods like tempeh and Soya sauce are produced by SSF alone

3. Some enzymes are produced by using SSF.

4. Leavening of bread by yeast done by SSF.

5. Mushroom cultivation on solid medium is an example for SSF

References

1. Aiba S, Humphrey A, E and Millis N.F. (1973), Biochemical engineering, Academic press, New York

2. Bailey J.E and Ollis, D.F (1986), Biochemical engineering fundamentals 2nd Edn, McGraw Hill, Inc New York.

3. Charles, M. (1985), "Fermentor design and scale up", comprehensive Biotechnology, Moo – Young, M.(Chief Ed) Vol 2, Pergamon Press.

4. Fogler, H.S. (1986), Elements of chemical Reaction engineering, Prentice Hall, Engle woods cliffs, N J.

5. L.E. Casida Jr. (1976), 'Industrial Microbiology' New Age international publications, New Delhi

6. Leven Spiel, O. (1974), Chemical Reaction engineering, Wiley Eastern Pvt. Ltd, New Delhi.

7. Michael L. Shuler and Fikret Kargi, (2002), Bioprocess Engineering basic concepts, Prentice Hall of India Pvt. Ltd., New Delhi.

8. Stanburry, P.F and Whitaker, A. (1993), Principles of Fermentation Technology, Pergamon Process Oxford.

9. Webb F.C (1964), Biochemical Engineering, D.Van Nostrand Co Ltd, London.

Review Questions

1. Briefly explain penicillin production.

2. How penicillin is recovered?

3. How citric acid is produced?

4. Write about citric acid recovery.

5. Explain production of Bakers yeast.

6. Describe production of single cell protein.

7. Write about vaccines.

8. What are enzymes?

9. Write some applications of enzymes.

10. Describe the production of HFCS.

11. Write about fermentation design.

12. What is air lift fermentor?

13. Write about tower fermentor.

14. Write about good design practice of fermentor.

15. Write about aseptic seals for fermentor.

16. What are aseptic operation of fermentor.

17. Write about aerobic fermentation.

18. Describe anaerobic fermentation.

19. Write about solid state fermentation.

20. Explain submerged fermentation.

CHAPTER 10

Aeration and Agitation in Bioprocess

Introduction

Aeration and agitation are very important operation in bioprocess units, basically to supply oxygen to the microorganisms and also very important for mixing the contents of fermentor.

Most of the fermentors are equipped with the agitators and the agitation causes the disintegration of the air bubbles and also turbulence in the reactor liquid.

The two important aspects considered during aeration and agitation of fermentation broths are the demand of oxygen by the microorganism and the supply of oxygen from the air bubbles to the fermentation liquid contents. Aeration and agitation provides oxygen for fermentation systems easily except in certain viscous fermentation broths i.e., fungal fermentations. Normally microbial oxygen demand varies with the variation dissolved oxygen concentration in the liquid. In bubble aeration, Power requirements in aeration factors effecting oxygen transfer coefficient, agitation and types of agitators will be briefly discussed here.

10.1 Mass Transfer in Cellular Systems

The general mass transfer of gas through the liquid to the microbial cell is shown in the Fig. 10.1

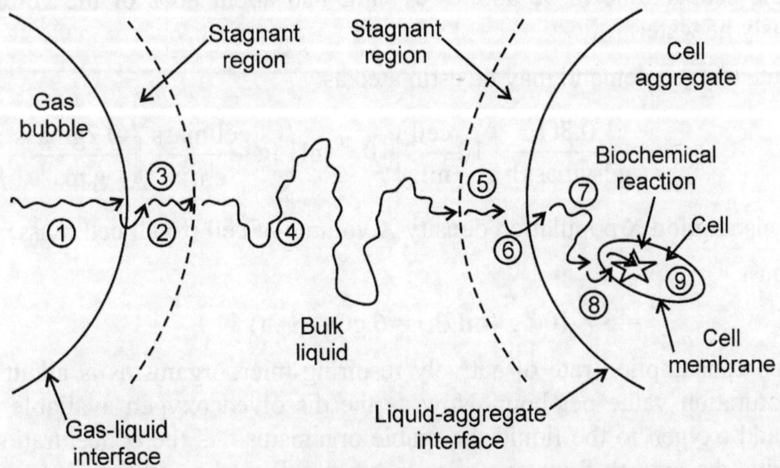

Fig. 10.1 Transfer oxygen from gas bubble in a cell.

The oxygen must pass through about '7' resistances as shown in the Fig. 10.1.

The resistances during the oxygen transport are given below,

1.   Diffusion from bulk gas to the gas - liquid interface
2.   Movement through the gas liquid interface.
3.   Diffusion of solute through the relatively unmixed liquid region adjacent to the bubble unto the well mixed bulk liquid
4.   Transport of solute through the bulk liquid to a second relatively unmixed liquid region surrounding the cells.
5.   Transport through the second unmixed liquid region associated with the cells.
6.   Diffusion transport into cellular floc, mycelia or soil particle.
7.   Transport across cell envelope and to intracellular reaction site.

The above given resistances are shown in the Fig. 10.1. The sixth resistance given above may disappear as the organism takes form of individual cell. The cells generally tend to gather at the gas – bubble liquid interface. So, the oxygen which is diffusing will pass through only one unmixed liquid region and not through bulk liquid region before reaching the cell. Here the bulk dissolved oxygen concentration will not represent the oxygen supply to the respiring microorganisms.

Mass Transfer Concepts (dissolved oxygen concentration in liquid)

The basic concept of hydro dynamics will be well understood by knowing the interactions between fluid motions and mass transfer. The solubility of oxygen at atmospheric pressure is about 10 ppm near ambient temperature. The approximate consumption of oxygen by yeast during respiration would be about 0.3 g of oxygen per gram of dry cell mass. The oxygen consumption at peak, for a population density of 10^9 cells per milliliter is estimated by considering the cell volume to be about 10^{-10} mL and about 80% of the volume assumed would obviously be water.

The absolute oxygen demand may be estimated as,

$$\frac{0.3O}{\text{gdrymass.h}}\left(10\frac{\text{cell}}{\text{mL}}\right)\left(10^{-10}\,\text{ml}\right)\left(1g\frac{\text{cellmass}}{\text{cm}^3}\right)\left(\frac{0.2\text{gmass}}{\text{g.mass}}\right)$$

Oxygen consumption X population density X volume of cell (totall cell mass) (1-0.8)

Rate by yeast

$$= 6 \times 10^{-3}\,\text{g/(mL.h)} = 6\,\text{gO}_2\,/(\text{L.h})$$

The oxygen consumption rate of actively respiring microorganisms is about '700' times the oxygen saturation value per hour. Now as the dissolved oxygen available is less, it is necessary to add oxygen to the liquid the viable organisms. As the concentration difference required to drive the oxygen from one zone to other is less, the addition oxygen to liquid is not an easy task. The equilibrium interfacial concentration of gas C_{gi} and liquid C_{li} of oxygen or hydrocarbons are related through a linear partition – law relationship such as Henry law may be given as,

$$M\, c_{li} = C_{gi} \qquad\qquad .....(1)$$

Considering that solute exchange rate across the interface is much larger than the bet transfer rate. Which is greatly in excess of the net flux for typical microbial consumption requirements. For example 1 atm of air and 25 °C, the oxygen collision rate at the surface is of the order of 10^{24} molecules per square cm per second.

At steady state, the oxygen transfer rate to the gas – liquid interface equals its transfer rate to the liquid film.

Now if C_g and C_i are oxygen concentrations in bulk gas and bulk liquid respectively, the transfer rate per gas side and liquid side can be written as,

$$\text{Oxygen flux } = \text{mol } O_2 \ (cm^2.s)$$
$$= K_g \,(C_g - C_{gi}) \text{ gas side} \qquad\qquad .....(2)$$
$$= K_l \,(C_{li} - C_l) \text{ liquid side}$$

The interfacial concentrations are not accessible in mass transfer measurements, over all mass transfer coefficient K_i and over all concentration driving force $C^{*}_1 - C_1$, is considered where $C^{*}_1 =$ liquid phase concentration at equilibrium with bulk gas phase.

$$Mc^{*}_1 = C_g \qquad\qquad .....(3)$$

So that solute flux is given by

$$\text{Flux} = K_i \,(C^{*}_1 - C_l) \qquad\qquad .....(4)$$

From equations 1, 2,3,and 4, the relation between overall mass – transfer coefficient K_l and physical parameters of the two film transport system K_g, K_l and M results

$$\frac{1}{K_e} = \frac{1}{k_e} + \frac{1}{Mk_g} \qquad\qquad .....(5)$$

M = much larger than unity for sparingly soluble substance and $K_g \gg K_l$ so, $K_c = K_l$ And all the resistance to mass transfer his on to liquid film side.

So the oxygen transfer rate per unit of reactor volume Qo_2 is given by

Qo_2 = oxygen absorption rate

$$= \frac{(\text{flux})(\text{interfacial volume})}{\text{Reactor liquid volume}}$$

$$= K_l \,(C_i^{*} - C_i) \, \frac{A}{V} \qquad\qquad .....(6)$$

$$= K_i \, a^1 \,(C_i^{*} - C_i)$$

Where a' = A/V is the gas p liquid interfacial are per unit liquid volume and the approximation $K_1 = k_1$. So k_1 will be used inplace of K_i. And a = gas – liquid interfacial area of per unit volume of bioreactor (gas + liquid) contents. Head space gas is excluded.

Qo_2 = local volumetric rate of o_2 consumption; the average volumetric rate of oxygen utilization (moles per time per volume) and $\overline{Q}_{o2}$ for an entire liquid volume V is given by

$$\overline{Q}_2 = \frac{1}{V}\int_0^V Q_{o2}\,dv \qquad .....(7)$$

$\overline{Q}_{o2} = Q_{o2}$ when interfacial area / volume and oxygen concentrations are uniform for entire vessel.

C_1^* is normally determined by temperature and composition of the medium. The influence of composition will become more complicated when liquid phase reaction is undergone by dissolved gas. For instance carbon dioxide may exist in liquid phase in any of four forms

1. CO_2 2. H_2CO_3 3. HCO_3 and 4. $CO^{2-}{}_3$. CO_2, H_2CO_3, $HCO^-{}_3$, $CO^{2-}{}_3$

The equilibrium relations (at 25 ^{0}C),

$$k_1 = \frac{[H^+][HCO_3^-]}{[CO_2]+[H_2CO_3]} = 10^{-6.3} \text{ M} \qquad .....(8)$$

$$K = \frac{[H^+][CO_3^{2-}]}{[HCO_3^-]} = 10^{-10..25} \text{ M} \qquad .....(9)$$

indicate that the total dissolved carbon concentration, C_T, as CO_2 is quite pH sensitive,

$$C_T = [CO_{2]}] + [H_2CO_3] + [HCO_3^-] +[CO_3^{2-}]$$

$$= [1+\frac{K_1}{H^+}+\frac{K_1K_2}{[H^+]^2}] \qquad .....(10)$$

This relation given in equation (10), show in Fig. 10.2.

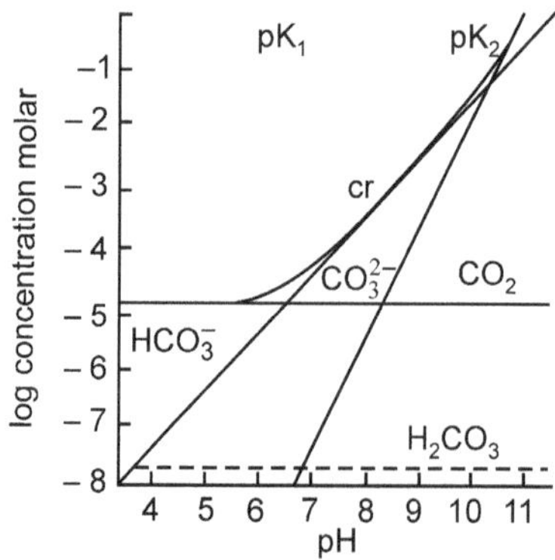

Fig. 10.2 Equilibrium concentration of dissolved CO_2, $HCO^-{}_3$, CO_3^- and H_2CO_3.

Fig. 10.2 shows that below pH 5, nearly all Carbon is dissolved molecular CO_2, while bicarbonate dominates when $7 < pH < 9$ and Carbonate for $pH>11$, only CO_2 (dissolved) is transported across gas liquid interface.

However, the coupling of reaction and mass transfer may occur under neutral to basic conditions

So the reversible reaction would be

Rapid

$$H_2CO_3 \leftrightarrow HCO_3 + H^+ \qquad\qquad .....(11)$$

And

$$K_{eq}\ (\text{at } 28^0C) = \frac{[H^+][HCO_3^-]}{[H_2CO_3]} = 2.5 \times 10^{-4} \qquad .....(12)$$

The important reaction,

$$H_2CO_3 \overset{K_1}{\underset{K_{-1}}{\leftrightarrow}} CO_2 + H_2O \qquad\qquad .....(13)$$

is slower with $k_1 = 20s^{-1}$ and $k_{-1} = 0.03\ s^-\ (25^0C)$.

So this CO_2 removal to gas phase may be chemical reaction or physical reaction [CO_2 dissolved to CO_2 gas]

10.2 Metabolic Oxygen Utilization Rates

The correlations frequently used during the design of aerobic biological reactors may be the slowest process step which is the oxygen transfer rate or the rate of cellular utilization of oxygen of other limiting substrate. Ofcourse the maximum possible mass transfer rate is found by setting $C_1 = 0$: and it is assumed that all oxygen entering the bulk solution is consumed rapidly. The oxygen utilization rate at its maximum would be $x\ \mu_{max}/Y_{o2}$. Where x = cell density; Y_{o2} = ratio of moles of cell carbon formed per mole of oxygen consumed .If $K_1\ a^1\ C_1^* >> x\ \mu_{max}/Y_{o2}$, the resistance to increased oxygen consumption is microbial metabolism and the reaction may be biochemical limited. However, the reverse inequality may lead to C_1 to near to zero and reactor seems to be mass transfer limited mode.

At steady state, the oxygen absorption and consumption rate should be balanced

Q_{o2} = absorption = consumption

$$K_1 a^1 (C_1^* - C_1) = \dfrac{X\mu}{Y_{O2}} \qquad .....(14)$$

The dependence of μ and C_1 assuming to be know eqn (14) may be used to evaluate C_1 and rate of oxygen utilization. And above some critical bulk oxygen concentration, the cell metabolic machinery is saturated with oxygen. Considering sufficient oxygen is available to accept immediately the entire cell would be rate limiting

$$Y_{O2} K_1 a^1 (C_1^* - C_1) = x\mu_{max} \dfrac{C_1}{K_{o2} + C_1} \qquad .....(15)$$

Here it is assumed that value of C_1 is much less than C_1^*, which is common phenomena in biological reactions.

So if $C_1 << C_1^*$ C_1 will be

$$C_L = C_L^* \left[\dfrac{\dfrac{Y_{O2} K_{O2} k_L a^1}{x\mu_{max}}}{\dfrac{1 - Y_{O2} C_1^* K_1 a^1}{x\mu_{max}}} \right] \qquad .....(16)$$

The C_1 evaluated may be greater than $C_T = 3\ K_{o2}$ (Critical oxygen value) then the rate of microbial utilization is limited by some other factor, e.g., low concentration of another substrate, even though the bulk solution has dissolved oxygen values below saturation values. The critical oxygen values for organism lie in the range of 0.003 to 0.05 mmol/L. Organisms like Penicillium molds have higher critical oxygen values and in such cases oxygen mass transfer is important.

K_{La}^1 values needed for process design, depends microbial oxygen demand $X\mu/Y_{o2}$.

Many factors that influence microbial oxygen demand are cell species, culture growth phase, Carbon, nutrients, pH, and the nature of the desired microbial process.

In Fig. 10.3 maximum specific O2 demand occurs in the early exponential phase through x greater after sometime. A peak value is the product $x\mu$ and thus the total oxygen demand occurs, near the end of the exponential phase and on set of stationary phase. Carbon nutrient affects oxygen demand. Glucose is metabolized more rapidly than the other carbohydrate.

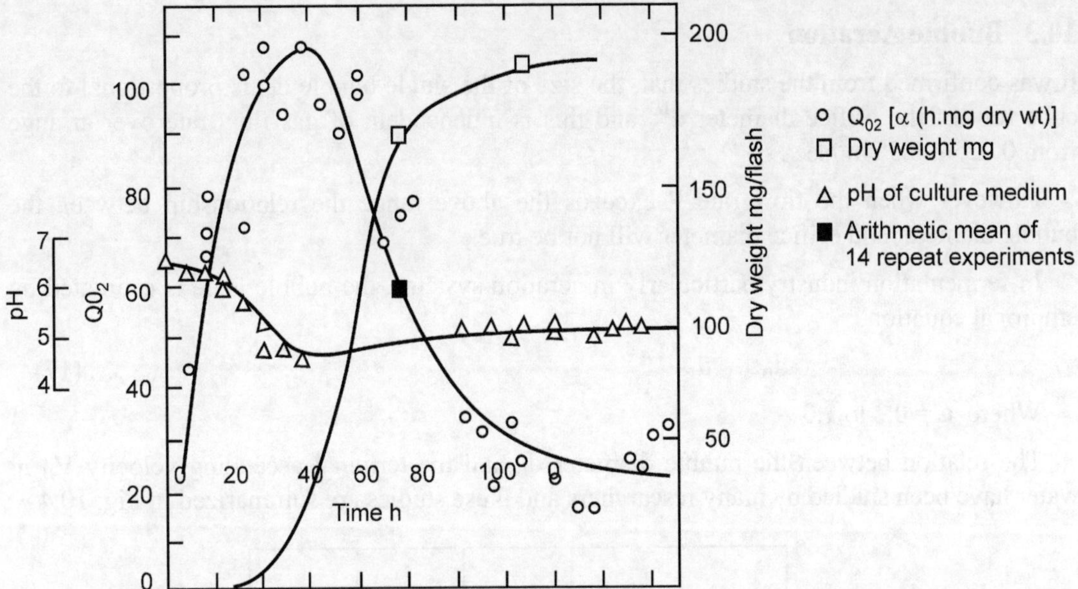

Fig. 10.3 Oxygen utilization rate in batch culture of *Nyrothacium verrucaria*.

Peak oxygen observed per Penicillium mold observed as given below Table 10.1

Table 10.1 Peak oxygen demand, for sugars.

Peak oxygen (mol/L.h.)	Sugar
4.9	Latose
6.7	Sucrose
13.4	Glucose

The oxygen utilized is useful per cell maintenance, respiratory oxidation for further growth (biosynthesis) and oxidation of substrates into metabolic end products.

Yield factor = moles of oxygen used per mole of carbon source metabolized. Y_{o2K}

Are given below (10.2)

Table 10.2 Yield factors for substrates.

Y_{o2}	Substrate
1.34	Methane
1.0	Paraffin
0.4	Carbohydrates

Oxygen may also be consumed in biotransformation

10.3 Bubble Aeration

It was confirmed from the studies that, the size of the single bubble d_B, is proportional to the cube root of the orifice diameter $d^{1/3}$ and this is independent of gas flow rate over arrange from 0.02 to 0.5 cm^3/sec .

However when the flow rate F exceeds the above range the relationship between the bubble diameter and orifice diameter will not be true.

In fermentation industry particularly in aeration systems, the bubble is d_B is estimated by empirical equation.

$$d_B \quad \alpha \quad F^\alpha \qquad \qquad .....(17)$$

Where $\alpha = 0.2$ to 1.0

The relation between the bubble diameter d_B and the terminal ascending velocity V_B in water have been studied by many researchers and these studies are summarized in Fig. 10.4

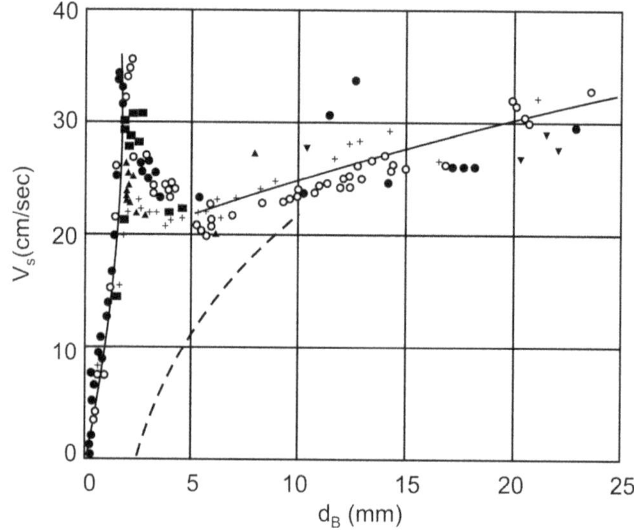

Fig. 10.4 Ascending terminal velocity V_s bubble diameter various symbols used in the indicate sources of original data.

Broken curve represents ascending velocity of CO_2 bubble in 1% aqueous polyox solution. (Pseudoplstic, non Newtonian liquid, consisting $K = 0.879 cm^{-1}$ flow behavior index $n = 0.540$ at 25^0C).

From the above Fig (10.4) it is evident that the broken curve, that the value of V_B was virtually zero unless d_B exceeds about 2.5 mm, the peculiar discontinuous region for the air water system vanished also v_B approached the value for air water system when d_B was larger than 10 mm.

Air bubbles used in the fermentation industry range process about 1.5 to 10 mm in diameter for single bubble the V_B values may range from 20 to 30 cm/sec in dilute fermentation media (similar to those of water)

10.4 Swarm of Bubbles

Normally we come across the swarm of bubble rather than single bubbles in fermentation industry. So we discuss about swarm of bubbles. The values of V_B for single bubbles approximately those of swarm of bubbles when there is no coalescence of bubbles in aeration and no appreciable distribution is found in bubble diameter. Most of the time aeration in fermentation industry is carried out by mechanical agitation, whereas, there is no mechanical agitation in air lift and tower fermentors. For mycelial cultivation, the physical properties of fermentation broth change significantly, as the process of fermentation proceeds. Also the motion of bubbles is affected by the turbulence caused by impellers and bubble diameter; either deforms or changes considerably during fermentation process. The V_B estimation by using Fig. 10.4 under above situations does not hold good.

So, for a fermentation process V_B can be estimated by using following equation per swarm of bubbles

$$V_B = \frac{FH_L}{H_0 V} \qquad \qquad .....(18)$$

Where

V_B = ascending velocity of bubbles in swarm

H_0 = hold up of bubbles

H_L = liquid depth

V = liquid volume

10.5 Aeration by Mechanical Means

The power number concept was developed by Rushton. Power requirements of liquid agitation for various types of impellers were measured by them. For ungassed and Newtonian liquids, the power Characteristics for impeller will be discussed here.

In agitated vessel velocity 'v' is proportional to tip velocity of the impeller

$$V \, \alpha \, n \, D_i \qquad \qquad .....(19)$$

Where n = rotation speed of the impeller

D_i = impeller diameter

The power number is defined as, the ratio, of external to inertial forces per unit volume

$$\text{Power number} = \frac{\text{External force}}{\text{Inertial force}} = \frac{\dfrac{pg_c}{nD_iD_i^3}}{\rho n^2 D_i}$$

$$= \frac{pg_c}{n^3 D_i^5 \rho} \qquad\qquad .....(20)$$

Where

P = Power requirements of agitation

G_c = conversion factor

Modified Reynolds number i.e., the ratio of inertial to viscous forces per unit volume depicts the liquid motions in agitated vessels.

$$\text{Modified Reynolds number} = \frac{\text{inertial force}}{\text{viscous force}} = \frac{nD_i^2 \gamma}{\mu} = N_{Re} \qquad ......(21)$$

Where μ = liquid viscosity

The power characteristics established by Rushton et.al., for liquid agitation with various types of impellers are shown Fig. 10.5.

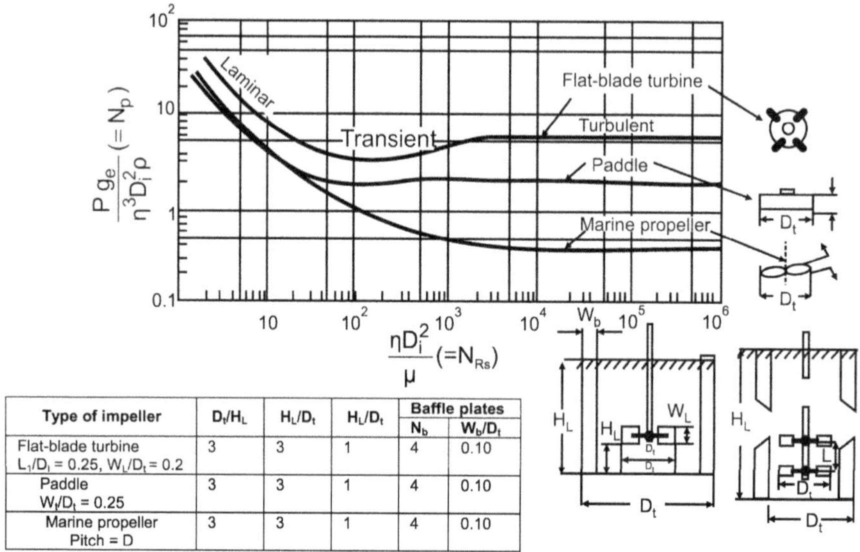

Type of impeller	D_t/H_L	H_L/D_t	H_L/D_t	Baffle plates	
				N_b	W_b/D_t
Flat-blade turbine L₁/D₁ = 0.25, W_L/D₁ = 0.2	3	3	1	4	0.10
Paddle W_i/D_t = 0.25	3	3	1	4	0.10
Marine propeller Pitch = D	3	3	1	4	0.10

Fig. 10.5 The plot of N_p Vs N_{Re}.

The power characteristics are shown by the power number, N_p and Modified Reynolds number, N_{Re} of single impeller on shaft. The geometrical ratios of impellers are given below the Fig. 10.5.

The values of "P" determined experimental by varying the values of n, D_i ρ and μ for each type of impeller series of geometrically similar systems are plotted in the Fig. 10.5. Most commonly used impeller in fermentation industry are shown in Fig. 10.5. The curves in the figure are plotted by taking certain geometrical ratios, however if the geometrically ratios are changed then the curves will be shifted for those that are plotted in Fig. 10.5. But nature of the each curve will remains same.

The standard geometrical ratios for flat blade turbine are shown in the in Fig. 10.5. For each type of impeller, the values of N_p at high values of N_{Re} remains constant.

10.6 Power Requirements in Aeration

In gassed systems the process requirements will be less than the power requirements of impeller in ungassed system. Most of the fermentation systems are gassed systems. The values of liquid density around the impeller will be less due to the systems of gas bubbles thus the power required for the gassed systems will be decreased. So if the bubble ascending through the liquid in the agitated tank does not effectively meet the impeller, then the power requirements for agitation will be affected slightly even in gassed systems.

Depending on the type of impeller and rate of aeration, the degree of power decrease in gassed systems to ungassed system Pg/P varies from 0.3 to 1.0. The term Pg/P will be correlated to degrees of dispersion of bubbles around the impeller and in the agitated tank. The following formula was presented by Ohyama etal.

$$Pg/p = f(N_a) \qquad\qquad .....(22)$$

$$Na = \frac{\dfrac{F}{D_1^2}}{nD_i} = \frac{\text{Apparebt velocity if gas (air) through sectional area of vessel}}{\text{Tip velocity of impeller}}$$

$$= F/nD_i^3$$

The aeration number 'N_a' would be useful for evaluating the degree of dispersion of bubble throughout the system.

Eqn (22) given the form of function 'f' for air – water system as per the experimental determination. The function 'f' also depends as type of impeller used as shown in Fig. 10.6.

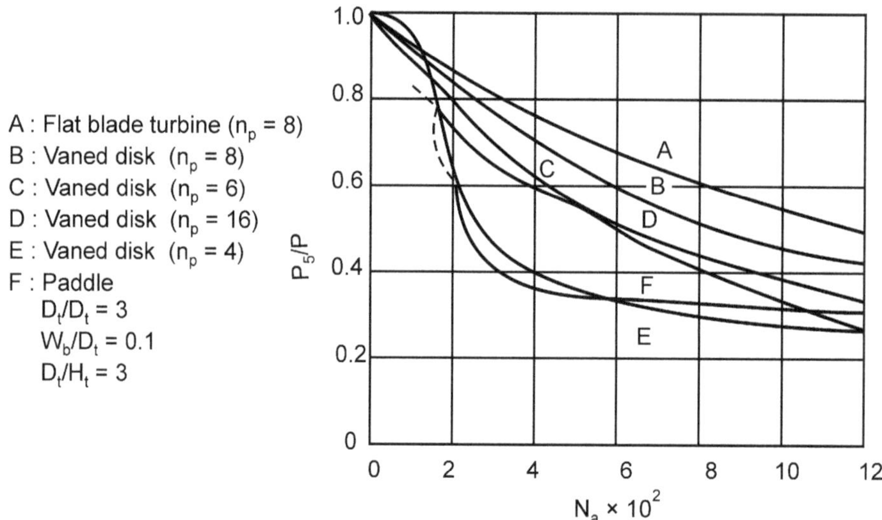

Fig. 10.6 Power requirements for agitation in a gassed system.

N_a = aeration number

The values of N_a varied for 0 to 12×10^{-3} in the experimental conducted while P_g/P varies for 1.0 to 0.3.

As per Michel et. al, the experimental correlation between P_g and operating variables, are given in equation

$$p_g \, \alpha \left(\frac{p^2 n D_i^3}{F^{0.56}} \right)^{0.45} \qquad\qquad .....(23)$$

The studies conducted were pertaining to values of P_g with standard flat blade turbine in air– water (liquid) system. The physical properties of liquids used are given below.

Liquid density $\rho = 0.8 - 1.65$ g/cm^3

Liquid density $\mu = 0.9 - 100$ cp

Surface tension $= 27 - 72$ dyne/cm

Equation (23) consider geometrically dissimilar systems also, while the physical properties of the liquids used did not explicitly affect the correlation with eqn (23), the system of aeration devices for ring sparger and open nozzle would be implicit.

Power Requirements in Non-newtonian Liquids

The average value of shear rate $(dv/dr)_{a,v}$ in air agitates vessel of non-Newton liquids according to Metzner et. al., was expressed as ,

$$(dv/dr)_{a.v} = \gamma.n \qquad\qquad .....(24)$$

r = proportionality constant which depends as the geometrical conditions to the agitated system and the type of non-Newtonian liquid.

n = rotation speed of the impeller τ

For non – Newtonian liquids

$$\mu_a = \frac{\tau g_c}{dv/dr} \qquad\qquad .....(25)$$

$$\tau g_c = f(dv/dr)$$

Provided μ_a = apparent viscosity of the non – Newtonian liquid

For a non – Newtonian liquid in an agitated vessel, the value of μ_a cannot be defined well, so the procedure proposed by Metzner et.al., for evaluating the value of the μ is:

(a) By varying the rotation speed n, of the impeller, measure the power requirement P, in an agitated vessel of laboratory scale filled with a non-Newtonian liquid then the values of the power number N_p are calculated.

(b) Replace the non-Newtonian liquid with a highly viscous Newtonian liquid whose values (ρ μ) are known, then using the same impeller and vessel, experimentally measure power consumed in agitating the liquid. For Newtonian liquid, determined the relation between the power number N_p and the modified Reynolds number N_{Re} Fig.10.6.

(c) The modified Reynolds number $n\, D_i^2 \rho / \mu_a$ incorporates the apparent viscosity, μ_a for each value of rotation speed n of the impeller it can be determined by equating both values of N_P the one calculated in accordance with step (a) above and other in accordance with step (b) using the relation between N_p and N_{Re} provided the lather values of $N_{Re} = n\, D_i^2 \rho / \mu_a$ are those of the Newtonian liquid examined as a reference. γ value of eqn(24) has to be determined.

a^1 values of μ_a determined in step C are substituted in eqn (25) to assess the values of dv/dr for each value of the provided the fermentation + (dv/dr) is given

b' By equating the values of dv/dr to $(dv/dr)_{ar}$ in eqn(24) the value of the γ is determined. once the value of γ is determined, 'P' value can be estimated as follows.

a" At a given rotation speed, n of the impeller the value of $(dv/dr)_{av}$ is determined with eqn (24) .

b" The value of μ_a is determined with eqn (25) followed by calculating of the Reynolds number, $nD_i^2 \rho / \mu_a$

c" with the correlation developed earlier between power number N_p, and modified Reynolds number N_{RE} for the Newtonian liquid the value of the N_p for the non-Newtonian liquid is determined for each value of modified Reynolds number, $nD_i^2 \rho / \mu_a$

In the above approximate procedure for calculating the power requirements for agitating a non-Newtonian liquid, the steps of equating the value $(dv/dr)_{av}$ in eqn (24) to that of dv/dr in equ (25) is for convenience only. The former value is related to the flow patterns of the non-Newtonian liquid in an agitated vessel. The latter is related to the value that assessed with viscometer.

Bubble Hold up in Aeration Vessel

Hold up of air bubble in vessels filled with water was studied by Richards et. al.

Hold up H_0 (%) is presented in abscissa of Fig. 10.7 while $(P/V)^{0.4} \times v_s^{0.5}$ is given is ordinate.

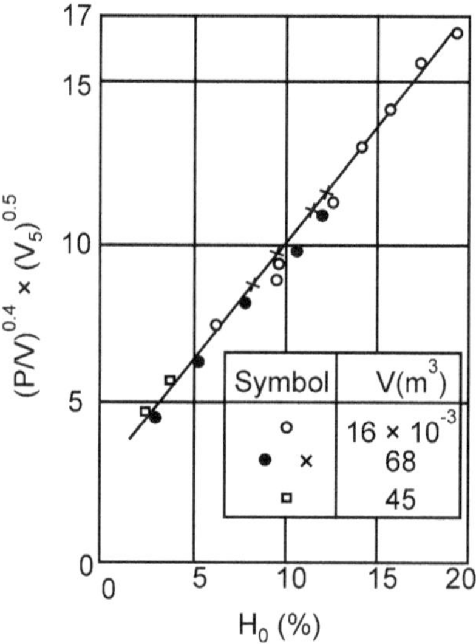

Fig. 10.7 Estimation of hold up, Ho, with power input per unit volume of ungassed liquid, P/V and nominal linear velocity v_s of air

The straight line drawn through the data present in Fig.10.7 approximation Calder banks correlation

10.7 Oxygen Transfer Coefficient and Operating Variables

Eckenfelder replotted the values of Reynolds number and Sherwood number using the K_L values published for bubbles rising through water. K_L values in the bubbles aeration of water γ were determined with the sulphite oxidation method by changing the water depth H_L. The bubble sizes in the experiments varied from $d_B = 0.5$ to 2.0 mm. The values of N_{sh} were dependent on the values of H_L, is primarily due to the experimental procedure of determining the values of K_L, which involves oxygen transfer both from the surface of bubbling water and in the transient state associated with the generation of bubbles. Due to these factors not clearly known K_L values are increased (and N_{Sh}) as the liquid depth H_L was lowered.

The effect of bubbles size from $d_B = 0.4$ to 8.8 mm as the K_L value size Fig. 10.8 of course the effects of transition state and bubbling water surface have been excluded.

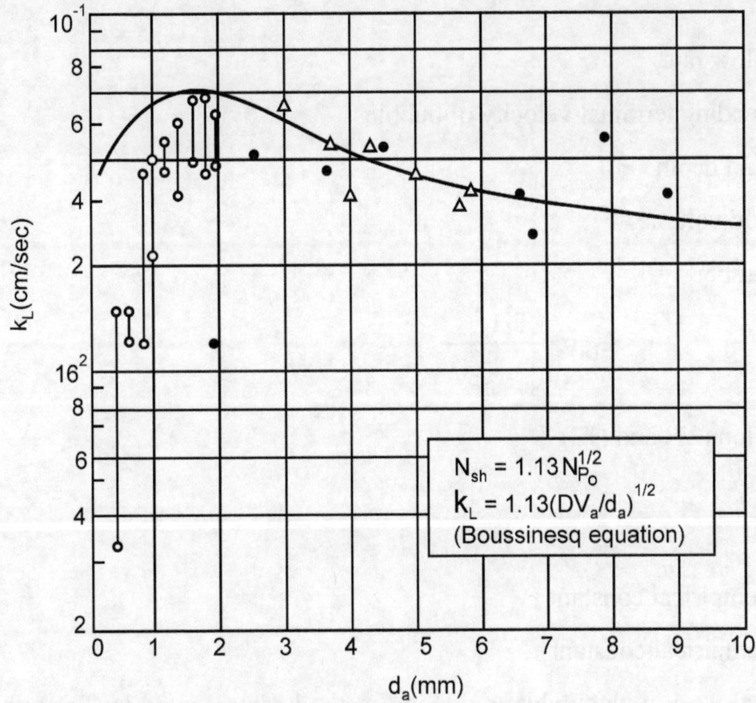

Fig. 10.8 Liquid (water)-film mass (oxygen)-transfer coefficient, k_L cm/sec. As effected by air bubble diameter d_B mm.

The data points given the Figure 10.8 mostly are represented by Boussinesq equation .For smaller bubbles anyhow the disagreement, observed between observation and calculation due to experimental errors or difficulties in experimentation. .

Calder bank et. al found that Boussinesq equation was applicable to predict K_L values for the following.

S.No.	CO_2	d_B
1.	90.6% aq. Glycol solution.	3 to 55 mm
2.	99.0% aq. Glycol solution	13 to 55 mm
3.	1.0% aq. polyox solution.	3 to 5 mm

Provided the coefficient 1.13 in Boussinesq equation was replaced by 0.65

The interfacial area 'a' between bubbles and liquid per unit of liquid is shown

$$a\alpha\dfrac{FH_L}{d_B v_B V} \qquad\qquad .....(26)$$

Where

F = air flow rate

r_B = ascending terminal velocity of bubbles

H_L = liquid depth

V = liquid volume

Schmidt number

$$K_{La}\alpha\dfrac{FH_L}{d_B^{1.5} v_B^{1/2} V} \qquad\qquad .....(27)$$

From equation (27) and (17)

We get

$$K_{La} V = \beta F^a \qquad\qquad .....(28)$$

Where β = empirical constant

$\acute{a}$ = empirical constant

Experimental work with sulphite oxidation method as presented by Eckenfelder supports the general form of eqn. 28 in Fig. 10.9

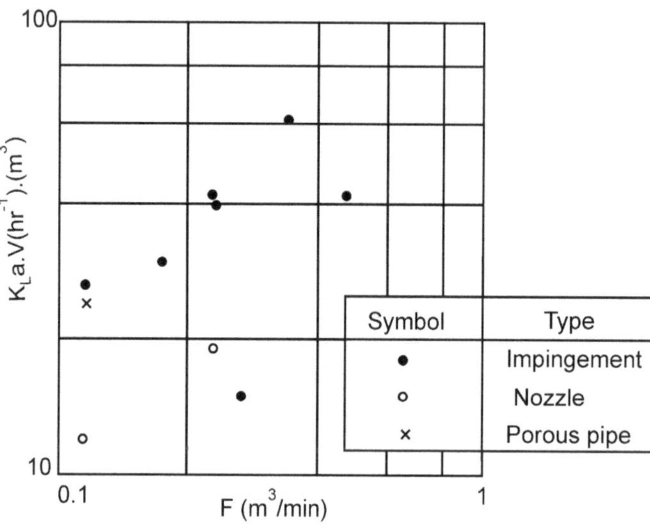

Fig. 10.9 Effect of aeration rate on the value of volumetric oxygen transfer coefficient.

k_{La}, times the volume of the aeration vessel, V. different aeration devices m used are given as different symbols.

10.8 Bubble Aeration and Mechanical Agitation

Dimensional analysis of the above problem shows,

$$K_L \ D_i/D \ \ \alpha \ \ (nD_i^2 \rho/\mu)^a \ (\mu \ /c_p T)^\tau$$

By keeping liquid viscosity μ, liquid density ρ, and molecular diffusivity (oxygen) D constant, eqn 27 can be simplified as,

$$K_L \ Di \ \alpha \ (nDi^2)$$

From literature,

$$\alpha = 0.5$$

Then $\qquad\qquad\qquad K_L \ \alpha \ \ n^{0.5}$ $\hspace{4cm}$ .....(30)

As per Calder bank, 'a' value per stirred vessel is

$$a\alpha \left\{ \frac{\left(\dfrac{P_g}{V}\right)^{0.4} \gamma^{0.2}}{\sigma^{0.6}} \right\} \left(\frac{v_s}{v_B}\right)^{0.5} \hspace{3cm} .....(31)$$

Now is ρ= liquid density σ = surface tension of liquid

V_B = ascending bubble velocity are constant

Eqn (31)

$$a \ \alpha \ \{ P_g/V\}^{0.4} \ V_s^{0.5} \ n^{0.5} \hspace{3cm} .....(32)$$

From eqn 30 to eqn 32, Richards et.al derived,

$$K_{La} \ \alpha \ \{p_g/v\}^{0.4} v_s^{0.5} n^{0.5} \hspace{3cm} .....(33)$$

They correlated their experimental data with eqn.33.

Experimental results of cooper et. al., presented in Fig 10.10.

It is interesting to note that from the Figure that identical values of absorption number result, irrespective of tank size, if the values of power consumed per unit volume of liquid are equated in a series of geometrically similar systems.

The value of K_V and other factors are taken in metric units to avoid inconsistency of units.

$$K_V = 0.0635 \ (P_g/V)^{0.95} \ V_s^{0.67} \hspace{3cm} .....(34)$$

K_V = volumetric oxygen transfer coefficient, K_g mole / hr.m^3 atm

P_g/V = power input per unit volume of liquid in gassed systems, HP/m^3

V_s = nominal (superficial) air velocity based on empty cross sectional area of vessel, m/dr

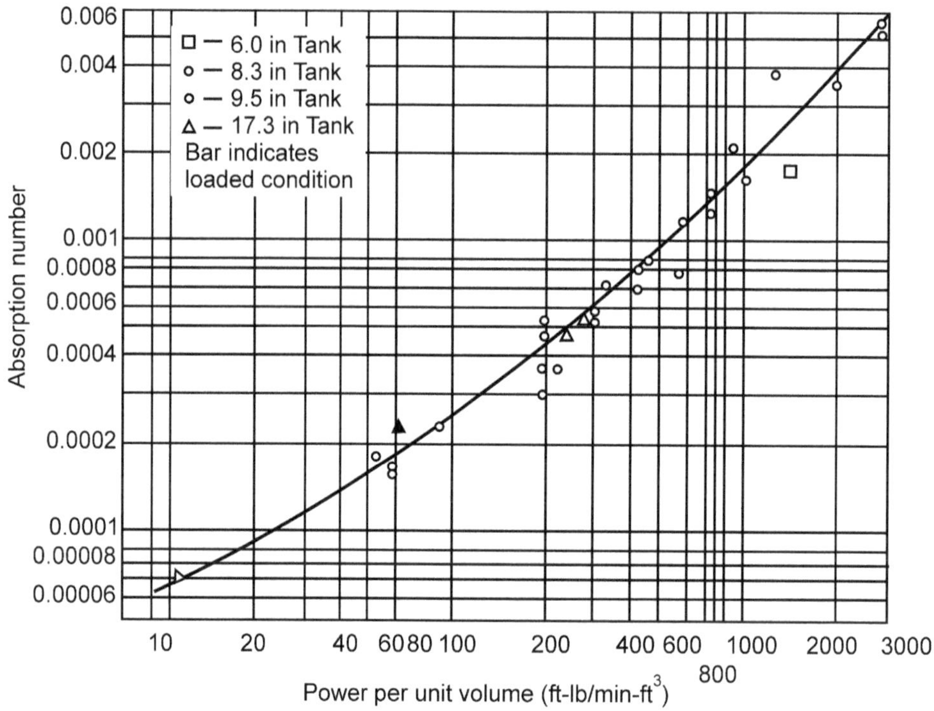

$$\text{Absorption number} = \frac{\text{Volumetric absorption coefficient}}{\left(\begin{array}{c}\text{nominal air velocity based on}\\\text{cross - sectional area of vessel}\end{array}\right)^{0.67}} = \frac{\left(\dfrac{\text{lb moles}}{\text{ft}^3 \text{ hr atm}}\right)}{\left(\dfrac{\text{ft}}{\text{hr}}\right)^{0.67}}$$

Fig. 10.10 Plot of absorption number versus power requirement
for agitation per unit volume of aerated liquid (water).

Equ (34) can be applied for the following conditions

    For 1 set of impeller $V_s < 90$ m/hr

    For 2 set of impeller $V_s < 150$ m/hr

Where

$$P_g/V > 0.1 \text{ HP/m}^3$$

$$\text{HL}/D_i = 1.0$$

For a paddle impeller

$$K_V = 0.038 \, (P_g/V)^{5.3} \, V_s^{0.67} \qquad\qquad\qquad .....(35)$$

Where

    $V_s < 21$ m/h

    $P_g/v > 0.06$ HP/m^3

    $H_L/D_t = 1.0$

Values of K_V for H_L/D_t = 2 to 4 are about 50% larger than those per H_L/D_t = 1.0, if the P_g/V is not changed

While designing fermentors if volumetric coefficient of oxygen transfer K_{ca} or K_v are not available

Equation (34) and Equation (35) or Fig. 10.10, may be used.

10.9 Other Factors Affecting the Values of Oxygen Transfer Coefficients

(a) *Temperature*

According to the analysis of 'O' Conner the K_{La} values at 30^0C and 10^0C are 15% larger and smaller than those at 20^0C respectively. This analysis is based on oxygen transfer during bubble aeration activated sludge process. Nevertheless the studies for fermentation may also be carried out.

(b) *Organic Substrate*

During bubble aeration of water the K_{La} values decreased to about one third of the initial value when 1% peptone was added. So the K_{La} will be about 40% of that without any organic substance.

Similarly when alcohol, Ketone and ester, added to distil water during bubble aeration. K_{La} value in distilled water increased by 50 – 100% with water the addition of these substances in water to the extent of 20 ppm in each case.

(c) *Surface Active agents*

The effect of addition of surface active agent on K, K_{La} during bubble aeration is shown in Fig. 10.11.

When sodium lauryl sulphite was added to water in small amounts, K_L was decreased sharply, However the value of K_{La} first decreased sharply the gradually recovered with the addition of surface active agent as shown in Fig. 10.11.

(d) *Mycelium*

The value of K_{La} will be affected by physical properties like viscosity μ or apparent viscosity μ_a and thin is applicable to mycelia fermentations Fig. 10.12.

The effect of mold concentration of $K_L a$ value is given in Fig. 10.12.

Newly prepared mycelia of A. niger is medium of sucrose, salts, and corn steep liquor in a fermentor. K_{La} was measured polarogarphically. The abscissa denotes the mycelium concentration by dry weight.

(e) *Type of Sparger*

The values of K_{La} various types of aerators found by sulphite oxidation method did not differ appreciably when compared on the basis of power consumption per unit of liquid. Vide Fig. 10.13.

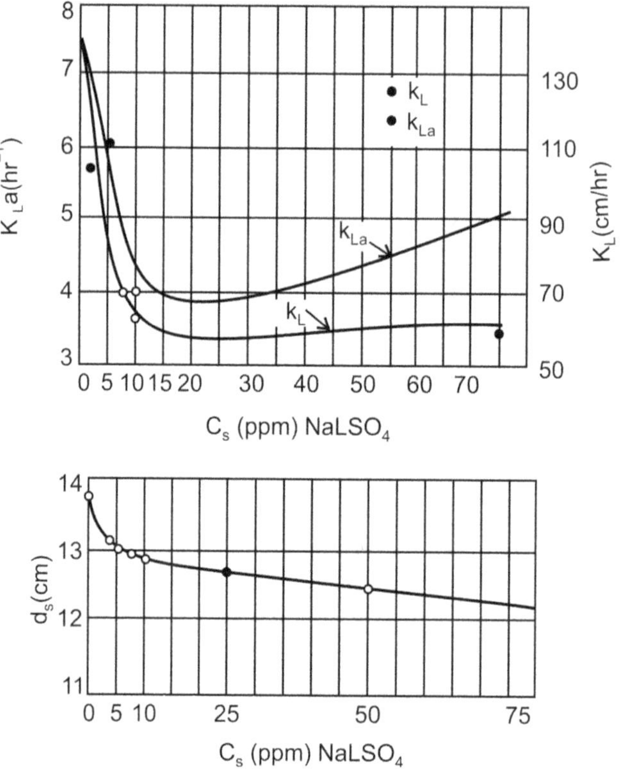

Fig. 10.11 Effect of surface active agent on k_{La} and k_L ; the upper section of the figure shows that the value of k_{La} is decreased by adding sodium lauryl sulfate (a surfactant); The lower section indicates the effect of surfactant on bubble size, d_B

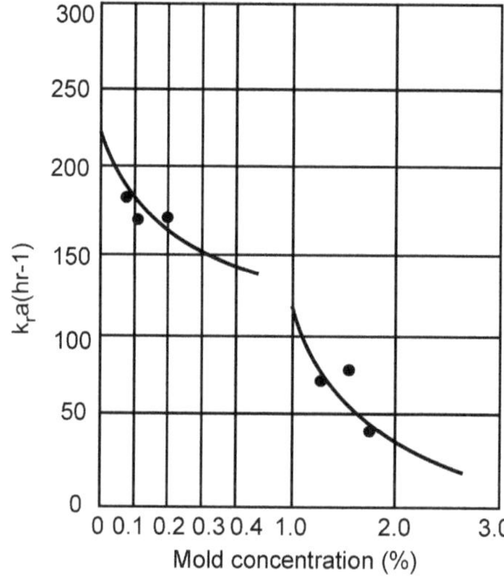

Fig. 10.12 Decrease of k_{La} values depending on the mycelial concentration in broth.

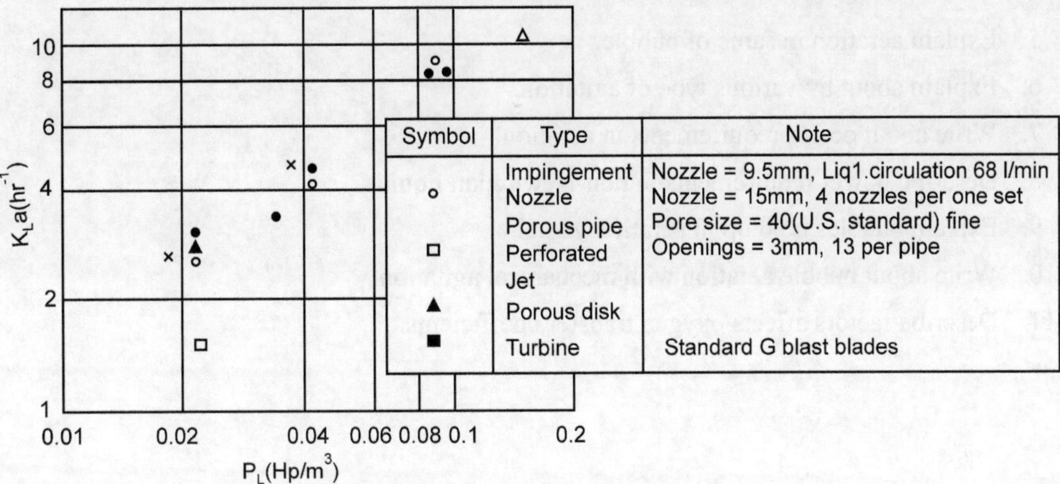

Fig. 10.13 Performance of various types of aeration devices.

The power requirement for all the spargers except for turbine are primarily the power consumed in sparing air with liquid.

References

1. Aiba, S. and Toda, K. (1963), "The effect of surface active agent on oxygen absorption in bubble aeration" J. Gen. Appl. Microbial.

2. Aiba. S, A.E. Humphrey and N.F. Millis (1973) "Biochemical engineering" Academic Press Inc, New York

3. Baily J.E and Olllis D.F (1986) "Biochemical engineering fundamentals". Mcgraw-Hill Book company, New York.

4. Calderbank, P.H. (1967), "Mass Transfer in Fermentation equipment" N. Blackebrough (Ed) Biochemical and Biological Engineering Science Vol 1, Academic Press Inc., New York.

5. Eckenfelder, W.W., Jr and E.L. Barnhar, (1961), "The effect of organic substance on the transfer of oxygen from bubbles into water" AlChE J.

6. Taguchi. H, and S. Miyamoto, (1966) "Power requirement in Non–Newtonian Fermentation broth", Biotech. Bioeng.

Review Questions

1. Write about mass transfer in cellular systems.

2. Describe basic mass transfer concept.

3. Explain about bubble aeration.

4. Write about bubble aeration.

5. Explain aeration in same of bubble.

6. Explain about by various type of agitation.

7. Write about power requirements in aeration.

8. Describe power requirements in non-Newtonian liquids.

9. Explain bubbles hold up in aeration vessel.

10. Write about bubble aeration with mechanical agitation.

11. Describe factors affects oxygen transfer coefficients.

Separation Process in Product Recovery

Separation plays an important role in product recovery and product purification in a bioprocess industry. It is essential to separate undesirable materials from final product, produced by a typical fermentation. Most of the undesirable material comprises mixture of cells, soluble extra cellular products, intracellular products, and unconverted substrate etc.

The costs of separations during recovery or purification of the bio-products are more than 50% of manufacturing cost.

The separation process to be adopted mainly depends on initial fermented broth characteristics like viscosity, industrial particles etc. Final product concentration and product form desired like, crystals or dry powders. The fermented broth will be subjected to various sequences of operations before final product. Some of these sequences are given below.

Sequence of separation process of fermented broth

In general, the separation techniques pertain to any typical bioprocess can be derived from categories 1. Removal of particulates 2. Primary isolation 3. Purification 4. Final product isolation.

Removal of Particulates

The separation of insoluble particulates like bio-mass, insoluble particles and macromolecules, from fermentation broth is achieved by following various methods. Some times the fermentation broth may be prepared before separation with coagulants and flocculants or pH may be adjusted. Various methods of separation may include centrifugation filtration, cell disruption and separation.

11.1 Centrifugation

The cells from the fermentation broth are removed by using centrifugation varies between 100 to 0.1 μm. Solid liquid separation in gravitational field will be considered before dealing centrifugation and sedimentation. In undressed settling, particle settles from high particle density suspension.

Solid particles settling in liquid by gravitation is mainly due to the force like, (F_a) gravitational force, (F_D) drag force and (F_B) buoyant force.

Forces acting on a particle balance each other when particle reach a terminal settling velocity and the net force would be zero.

Sequence of separation process of fermented broth

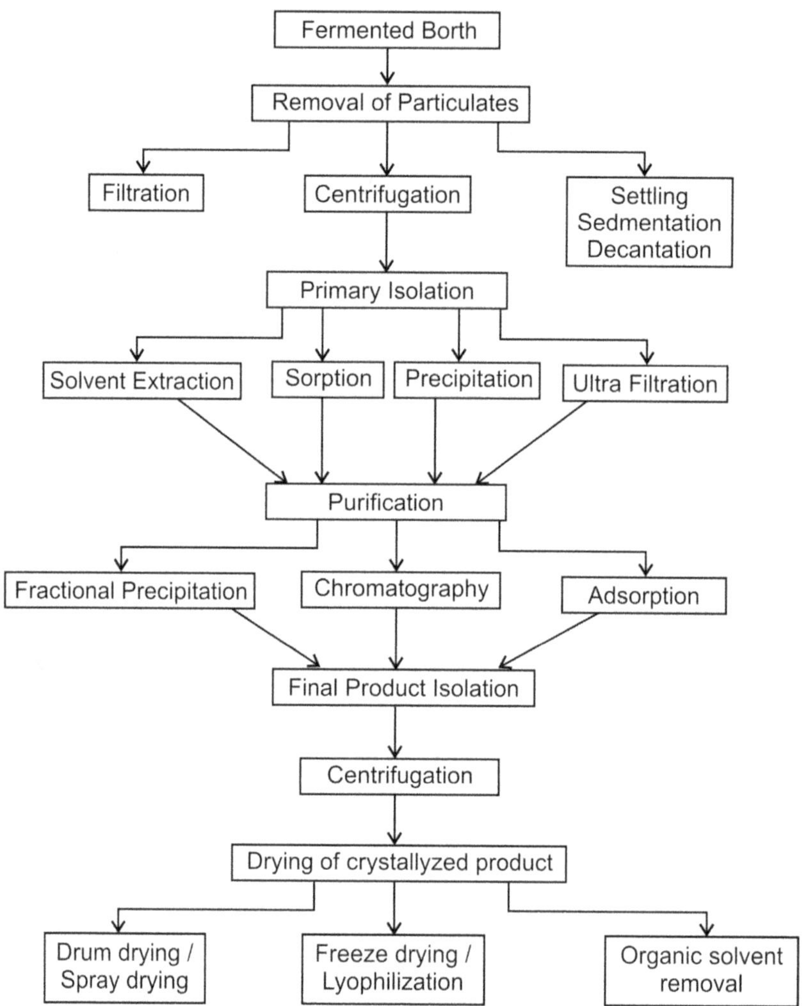

So , $\qquad F_G = F_D + F_B$ .....(1)

Where

$$F_G = \pi/6 \; D^3_\gamma \rho_p \; g/g_c \qquad\qquad .....(2)$$

$$F_D = \pi/6 \; D^3_P \rho_f \; g/g_i \qquad\qquad .....(3)$$

$$F_B = \pi/6 \; \rho_P U_o^2 A \qquad\qquad .....(4)$$

C_D = drag coefficient ρ_f = fluid velocity

= relative velocity between the fluid and particle (or) terminal velocity of particle

A = cross sectional area of the particles perpendicular to the direction of fluid flow. Ex : for a sphere $A = (\pi/4) \; D_P^2$

F_D = drag force exerted by fluid on solid particles

Now for spherical particles, when $Re_P < 0.3$, the drag force F_D is given by stokes equation

$$F_D = 3 \pi \mu \; D_P \; \mu_0 \; \frac{1}{g_c} \qquad\qquad .....(5)$$

Upon substituting (5) in eqn (4) we have

$$C_D = \frac{24}{R_{eP}} \qquad\qquad .....(6)$$

If Re_P is from 1 to 10,000 discharge coefficient C_D may be given by

$$C_D = \frac{24}{Re\,p} + \frac{3}{\sqrt{Re\,p}} + 0.34 \qquad\qquad .....(7)$$

Where $Re = D_p U_o \rho_f /\mu$

Upon substituting eqn (2) , (3), (4) in eqn (1) we get

$$3\pi \, \mu \, D_0 \, \mu_0 \; = \; \frac{\pi}{6} D_p^3 (\rho_p - \rho_f) g \qquad\qquad .....(8)$$

$$\mu_0 = \frac{g D_p^2 (\rho - \rho_{f_[})}{18\mu} \qquad\qquad .....(9)$$

Where

D_P = particle diameter

ρ_p = particle density

The terminal separation of particle μ_{oc} in a centrifugal field is given as

$$\mu_{0c} = \frac{rW^2 D_p^2 (\rho - \rho_{f[})}{18\mu} \qquad .....(10)$$

This is obtained by substituting centrifugal acceleration for gravitational acceleration in equ (9)

OR $$\mu_{0c} = \frac{gzD_p^2 (\rho - \rho_{f[})}{18\mu} = ZU_0 \qquad .....(11)$$

Where,

$Z = rw^2/g = $ centrifugal factor

$r = $ radial distance from the central axis of rotation

$w = $ angular velocity ($w = 2\pi Nr$)

The above equations are valid for dilute solution where particle – particle interaction can be neglected. When particle concentration are high in fluid, particle agglomerate from swarms and terminal velocity reduces for

$$\mu \, D_P \, U \, (1 + \beta_o \, \frac{D_1}{L}) \qquad .....(12)$$

Where $\mu = $ terminal velocity of particles under hindered settling

$L = $ Average distance between adjacent particles

$\beta_0 = $ Hindered settling coefficient

β_0 is 1.6 for rectangular arrangement of particles

In dilute solution coefficient $D_P/L << 1$, F^1_D will approach F_D. Hindered settling coefficient β_0 is a function of volume fraction of particles α and shape of particles,

$$\frac{\mu_c}{\mu_{oc}} = \frac{1}{1 + \alpha^1 \alpha^{1/3}} \qquad .....(13)$$

$\acute{\alpha} = $ empirically correlated with α and depends on particles shape also

$\mu_c = $ terminal velocity of particle of a given size and shape in crystal fields under hindered settling conditions

Capacity of centrifuge to handle a given flow rate of broth is considered while designing a centrifuge. Two popular centrifuges, Bowl or tubular bowl are shown in Fig. 11.1.

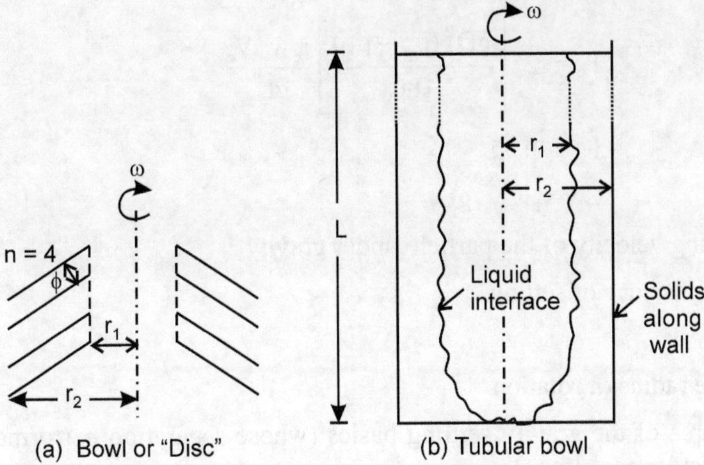

Fig. 11.1 Common centrifuges of two types are shown (a) Disc or bowl centrifuge with multiple separator Bowls or discs (b) A tubular centrifuge.

For discs or bowl centrifuge, scale up factor depends on two radial distances, small r_1 and r_2, and angle Φ the rotational speed (ω) and n the number discs or effective separator bowls. The tubular bowl centrifuge can operate with continuous liquid feed with span of continuous operation determinates by the capacity of bowl to collect solids. Where L = the height of unit.

The design consideration of a centrifuge, comprises,

1. The capacity of centrifuge to handle a given flow rate of fermentation broth

2. The velocity of the particle in the centrifugal fluid

3. The travel distance of the particle

Now consider distance, of travel of particle in one dimension case

$$Y = U_{oC}\, t \qquad\qquad \ldots\ldots(14)$$

Substituting U_{oC} from eqn (11) into eqn (14)

And the time of insert is $t = V_c/F_c$

We have

$$y = \frac{rw^2 D_p^2 (\rho_p - \rho_f)}{18\mu} \frac{V_c}{F_c} \qquad\qquad \ldots\ldots(15)$$

Where

V_c = liquid volume in the centrifuge

F_c = liquid flow rate through two centrifuge

Substituting $y = L_c/Z$ and $r = r_e$ in to eqn (15)

We get

$$F_c = 2 \left[\frac{gD_p^2(\rho_p - \rho_f)}{18\mu} \right] \frac{r_e w^2 V_c}{gL_c} \qquad .....(16)$$

OR

$$F_C = 2U_0\Sigma \qquad .....(17)$$

$$\Sigma = v_e w^2 v_e/gL_e$$

U_0 = free setting velocity of the particle under gravity

L_c = effective distance of setting

$\quad = 2y$

R_e = effective radius of rotation

Σ = surface area of the gravity settling basics (whose separation performance is equal to that of the a particular centrifuge)

For a tubular centrifuge

$$\varepsilon = \frac{2\pi L w^2}{g} \left(\frac{3}{4} r_2^2 + \frac{1}{4} r_1^2 \right) \qquad .....(18)$$

L = length of cylindrical separator

V_1 = minimum radius centrifugal distance

R_2 = outer radius of centrifugal distance

For Bowl centrifuge Σ is

$$\varepsilon = \frac{2\pi n^1 w^2}{3g \tan \theta^1} (r_2^3 - r_1^3) \qquad .....(19)$$

Where

N^1 = number of separation bowls

Φ^1 = angle of inclination of side walls

Now assuring that force settling velocities of particles are same.

For scale up centrifugal the following equate may be used

$$\frac{F_{c2}}{F_{c1}} = \frac{\varepsilon_2}{\varepsilon_1} \qquad .....(20)$$

F_{c1} = Flow rate through small centrifuge

F_{c2} = Flow rate through large centrifuge

11.2 Filtration

Filtration is frequently used to remove suspended biomass from the fermented broth. Plate and frame type filter may be used for small fermentations while, continuous rotary vacuum filter may be used for handling large volume of fermentation broth. Penicillin mycelia are removed by using rotary vacuum filter with steering discharge. But streptomyces mycelia are removed by using precoated filter cloth filter aid like

Diatomaceous earth and cake is removed by using knife blade.

The drum of the rotary vacuum filter is covered with a pre-coat prior filtration; coagulated agents like alum may be added to the liquid before sending to the filter. Vacuum is applied to the drum as the drum rotates the microbial cells adhere to the drum surface. The layer of cells increases over a period of time and as the drum rotates the knife cuts the cake.

Mycelium for the fermentation broth is removed by rotary vacuum filters.

The rate of filtration for constant pressure filtration determined by resistance of filter medium and cells

$$\frac{dv}{dt} = \frac{g_c \Delta p A}{(r_m + r_c)\mu} \qquad \qquad .....(21)$$

Where

V = volume of filtrate

A = surface area of the filter

ΔP = pressure drop through the cake and filter medium

μ = viscosity of filtrate

r_m = resistance of cake

A schematic diagram rotary vacuum filter in given Fig. 11.2.

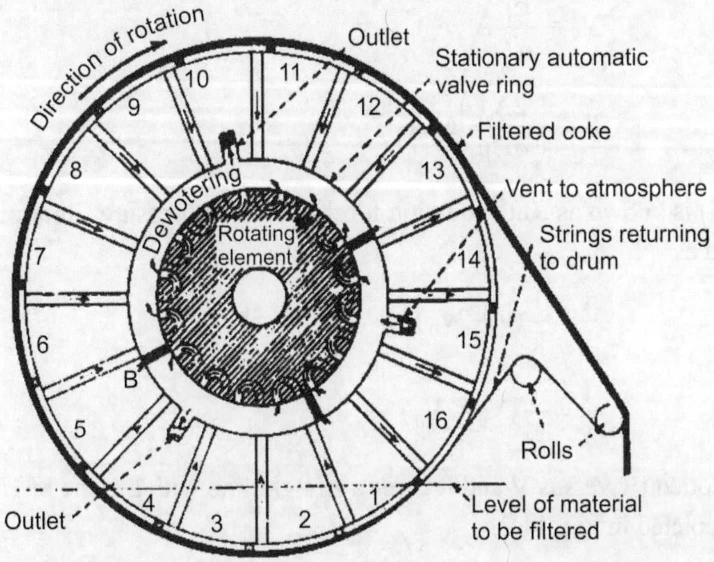

Fig. 11.2 Schematic diagram of a string filter in operation.

The resistance of filter medium r_m depends on characteristics of filter medium

During filtration the cake resistance r_c increases ad after passing some time,

$$r_c > r_m$$

$$r_c = \alpha \frac{N}{A} = \alpha \frac{\rho v}{A} \qquad .....(22)$$

where

W = Total weight of cake

C = the weight of the cake deposits per unit volume of filtrate

α = Average specific resistance of cake

So the total weight of cake, would be,

$$W = CV \qquad .....(23)$$

Now substituting equ (22) and (23) in equ (21) we get

$$.....(24)$$

Upon integrating of equ (24) (v to v and t, 0 to t)

$$V^2 = 2VV_0 = Kt \qquad .....(25)$$

Where

$$V_0 = \frac{r_m}{\alpha \ell} A$$

$$K = \left(\frac{2A^2}{\alpha C \mu} \right) \Delta p.g_c$$

Equation (25) is known as Ruth equation for the constant pressure filtration. This equation may be arranged as

$$\frac{t}{v} = \frac{1}{k}(v + 2v_0) \qquad .....(26)$$

$$\frac{t}{v} = \frac{1}{K}(V + 2V_0)$$

The graph plotted t/v versus V and it gives a straight line with a slope of 1/K and intercept of 2 V_0/K , as depicted in Fig. 11.3.

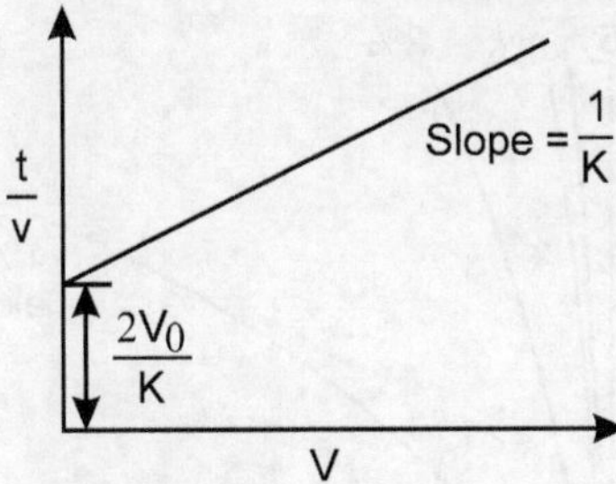

Fig. 11.3 Plot of t/V V$_s$ V

R_m and α can be calculated from experimentally determined values of K and V_0

The drum rotates at a constant speed (n rps) and some Portion of drum surface area is immersed in suspension (Φ) in trough. The filtration time period is Φ/n per revolution of drum.

Equation (25) can be written as

$$\left(\frac{V^1}{n}\right)^2 + 2\frac{V^1}{n}V_0 = K\frac{\phi}{n} \qquad .....(27)$$

Where

V^1 = filtrate volume per unit time (volume/ time)

V^1/X = volume of filtrate (filtered) for one revolution of drum

The assumptions made during the filtration include, cake formed is an incompressible cake with a constant specific resistance. But fermentation cakes are generally compressible, and α various with ΔP. Specific cake resistance also depends on filter aid concentration (1% to 5%). The specifies cake resistances decreases as the concentration of filter aid increases as shown Fig. 11.4.

The parameters like, pH, Viscosity, and Composition of the medium will affect the specific resistance of cake. pH effect or cake resistance is shown in the Fig (11.5) When pH decreases specific resistance also decreases.

The filterability of the fermentation broth of penicillin fermentation improves by heat treating the mycelia protein to coagulate.

The filterability and antibiotic activity is affected by period of fermentation. The cake resistance and antibiotic activity are minimized by using penicillin fermentation time of 180 – 200 hr. The conditions required for streptomycin fermentation broth are (1) pH = 3.6 filter aid 2 to 3% and heat treatment 30 to 60 min T = 80°C to 90°C. However micro porous filtration or ultra filtration can also be used for the purpose of cell separation.

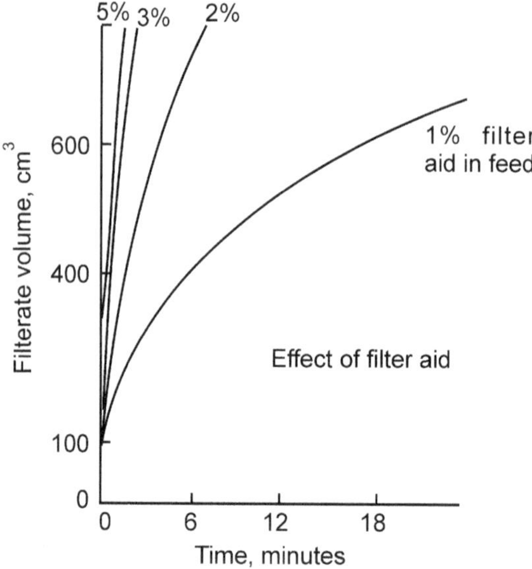

Fig. 11.4 Effect of filtration aid on specific resistance to filtration.

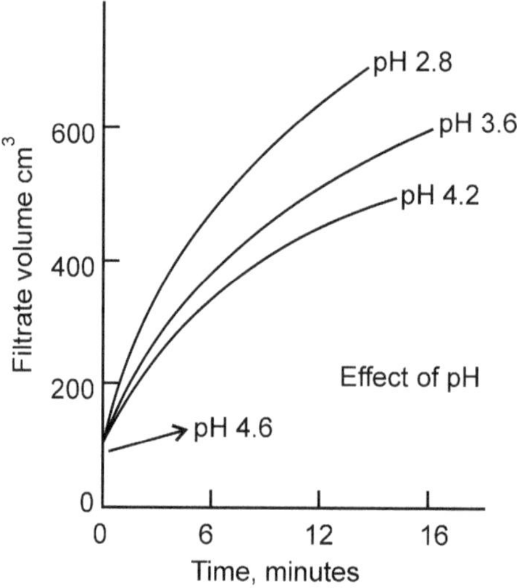

Fig. 11.5 Effect of pH on the rate of *Streptomyces griseus.*

11.3 Sedimentation

The cells can be separated by sedimentation if they agglomerate after the addition of coagulants like polyvalent ions. The suspension of particles settle slowly when concentration is high, unlike single particle settling which is independent of concentration. The settling velocity of such particles can be given by

$$\mu_s = k\, c^{-m} \qquad \text{where } m = 1.7 \text{ to } 2.6 \qquad\qquad .....(28)$$

The concentration of sludge occurs at a negligible rate when the cell concentration attains a maximum value C_{max}. The example for this phenomenon is settling of activated sludge having mixed culture bacteria. The sludge concentration can further be increased either by filtration or by centrifugation. An appropriate process for dewatering of sludge can be determined by thoroughly examining the various process available like, sedimentation, centrifugation, filtration and drying. Sedimentation mostly removes free water. Sedimentation rates may be improved in inclined tubes or narrow channels due to secondary flow phenomena. This results in evolution of clear zone rapidly at the top of inclined tube or channel.

11.4 Disruption of Cells

The cells are required to be disrupted when the final product required is intracellular, and have been separated for the liquid fermentation broth. Basically the cell disruption can be achieved either by mechanical or non-mechanical means.

(a) *Mechanical methods*

The cell disruption can be applied to liquid or solid.

Sonication

These ultra sonic vibrations functions by disrupting membrane of bacterial cells and the cell wall of bacterial cells. About 20 kc/s of wave density is used for this purpose. The gram negative bacterial cells are broken more easily than the gram positive cells and rod shaped bacteria are broken more easily than the bacterial cocci.

Ultrasonic waves are produced by an electronic generator and they are further converted into mechanical oscillations by a transducer i.e., titanium probes immersed in cell suspension. Cell disruption releases intracellular compounds like, enzymes and metabolites into the fermentation broth. Sonication or ultrasonic disruption may amount to denaturation of sensitive enzymes and cell debris fragmentation. If large volumes of broth are subjected to the sonication, heat may be dissipated during cell disintegration.

French press and the Gaulin–Manton press, function well at a laboratory scale. In French press the cell paste is filled in a hollow cylinder in a stainless–steel block and high pressure is applied. The cells disrupt as they are extruded through the value to atmospheric pressure as the cylinder has a needle valve at base. In continuous operation 'Ribi' fractionators similar to French press functions satisfactorily. But this requires pre-concentrated slurry or paste. The capacities up to 40,000 l/h can be handled in Rannie high pressure operating at 1000 bar. Some mechanical homogenizers contain small beads of 20 to 50 mesh size. One of such mechanical homogenizers is Dyne-Mill which is used at large scales. The size of Dyne-mill will be about 275 l and normally processes 340 kgdw/h of yeast or cell suspension of about 2000kg/hr.

Microorganisms like algae, bacteria, and fungi will be subjected to this type of cell disruptions. The horizontal chamber filled with 80% of beads is used for disrupting the cells of the liquid suspension. A shaft containing specially designed discs is fitted inside the grinding chamber. The cells will break due to high shearing and impact forces of numerous beads when rotated at high speeds. The cells which are broken will be discharged, the mechanical homogenizer having beads are advantageous due to broth temperature control than the devices that apply pressure. Pathogenic or genetically modified cells will be presented from release of organisms in aerosol since this device is totally enclosed.

Different processing devices are available for handling solid medium like frozen cell paste or cells attached to or with a solid matrix. Mostly ball mills are used in large scale processes. Hughes and X -presses are also used for cell disruption. A split block with a half – cylinder hollowed in each face is fitted in Hughes press. The frozen cell paste is placed in the hollow and pressure is applied to it from the cylinder into channels cut in the block. While in the X-press, at low temperature and pressure, the frozen cells are forced continuously through a small hole in a disc between two cylinders. The cell disruption occurs due to deformation of organisms embedded in the ice. In X- press the pressure levels up to several hundreds bars can be applied.

However the products may not be free of damages, when above methods are followed for cell disruption. Product quality is an important aspect which should not be ignited while selecting method for cell disruption.

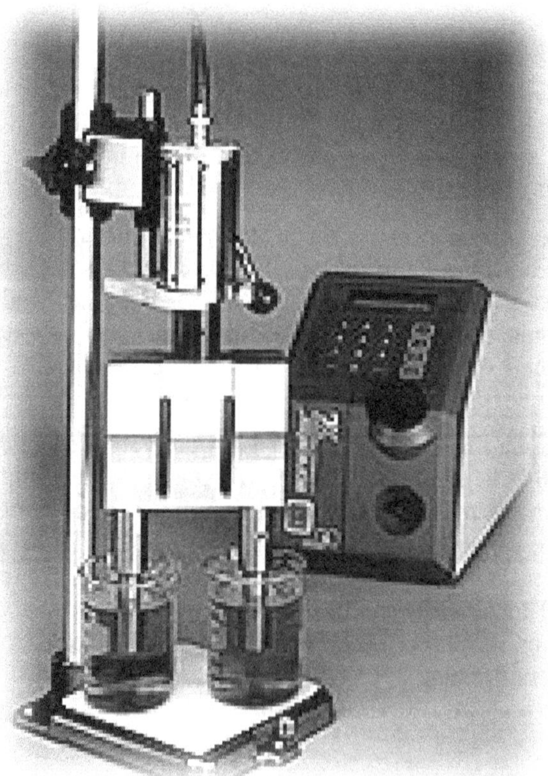

Fig. 11.7 Cell Disruption Methods-sonicator.

Non-mechanical Methods

The general methods used are applying osmotic shock and rupturing with ice crystals. The enzymes may be released into the media by freezing and thawing a cell paste when the cell wall and membranes will be broken. The release of enzymes may also take place by changing the osmotic pressure of the medium. Cell wall disruption may occur by treating the cells with acetone, butanol and buffers and then selecting to freeze drying.

Bacteria cell wall may be lysed by using enzymes like lysozyme. Mostly gram positive bacteria are lysed enzymatically than the gram negative bacteria. Cells may be preheated with EDTA (ethylene diamine tetra acetic acid) and subjected to lysozyme treatment. Since enzymatic hydrolysis is an expensive method, it is not frequently used in industry. Cells after disruption and releasing products into medium will be separated either by ultra centrifugation or ultra-filtration.

Primary Isolation

The primary isolation consists of extraction, precipitation, sorption, ultra-filtration.

11.5 Extraction Liquid – Liquid

The inhibitory fermentation products such as ethanol and – acetone butanol are generally extracted by liquid extraction. Antibiotics are also extracted by liquid extraction (using amyl acetate or iso-amyl acetate). Liquid extractant selected should normally be non–toxic, selective, immiscible and inexpensive with fermentation broth and it should have a high distribution coefficient for the product to be separated.

The solubility difference of this compound in one phase to other is an important criterion , during the extraction of one compound from one phase to other. The ratio of concentration of a compound distributed into phases of two immiscible liquids is known as the distribution coefficient.

$$K_D = \frac{Y_L}{X_H} \qquad \qquad .....(28)$$

Where

Y_L = concentration of soluble in light phase

X_H = concentration of solute in heavy phase

In most of the situations light phase would be a solvent while the heavy phase would be aqueous fermentation broth.

Considering K_D as constant and the solvents are immiscible (i.e., the mass flows of light and heavy phases are conserved, and $L_o - L_1 = L$ and $H_0 = H_1 = H$), a mass balance on solute extracted gives, Fig. 11.8.

$$H (X_0 - X_1) = L\, Y_1 \qquad \qquad .....(29)$$

$$\text{Or} \qquad \qquad X_1 = X_0 - \frac{L}{H} Y_1 \qquad \qquad .....(30)$$

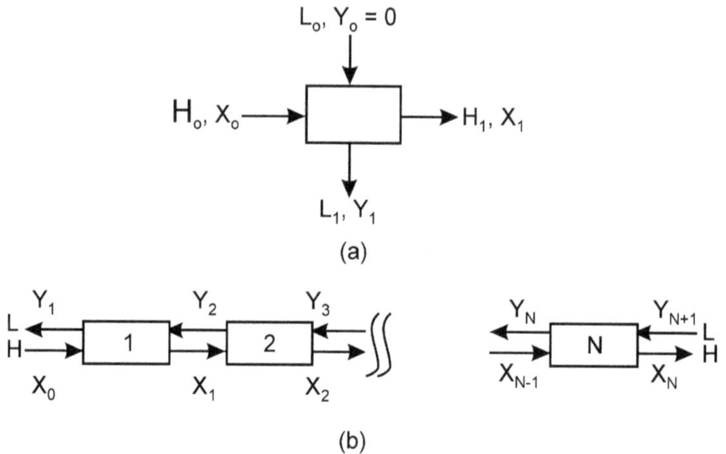

Fig. 11.8 (a) single stage (b) multi stage counter current extraction system.

As $K_D = Y_1/X_1$ eqn (30) can be

With as

$$X_1 = X_0 - \frac{LK_D}{H} X_1 \qquad \qquad .....(31)$$

$$\frac{X_1}{X_0} = \frac{1}{1 + (LK_D / H)} = \frac{1}{1 + E} \qquad \qquad .....(32)$$

Where

$E = (L\,K_0) / H$ is the extraction factor

A material balance on solute extracted for a counter current operation gives

$$R = \frac{E(E^n - 1)}{E - 1} \qquad \qquad .....(33)$$

Where

R = rejection ratio = (Weight of solute leaving in light phase) / (weight of solute leaving in heaving phase)

n = number of equilibrium stages

The fraction of solute extracted will be

$$\% \text{ extraction} = 1 - \frac{1}{R + 1} \qquad \qquad .....(34)$$

The reflection ratio, when the solute enters the system in light phase

$$R = \frac{E^n(E-1)}{E^n - 1} \qquad\qquad .....(35)$$

The relationship between E, X_M/X_o and n duplicated in Fig. 11.9.

This Fig (11.9) is used to determined number of stages (n) required for a certain degree of extraction (X_n/X_o) for given system (E).

The antibiotics are extracted by using solvents like amyl acetate or iso-amyl acetate from the fermentation broth.

Podbielneak extractors (continuous centrifugal) are commonly used for extracting antibiotics. The solubility of penicillin at pH is 2 to 3 is more in organic phase and solubility of penicillin is more at pH 8 to 9 in aqueous phase. So the extraction can be carried out many times by shifting pH in organic and aqueous phases to get pure product Fig. 11.10 gives the variation of distributions coefficient (solvent aqueous) with pH in penicillin extraction by solvents like amyl acetate

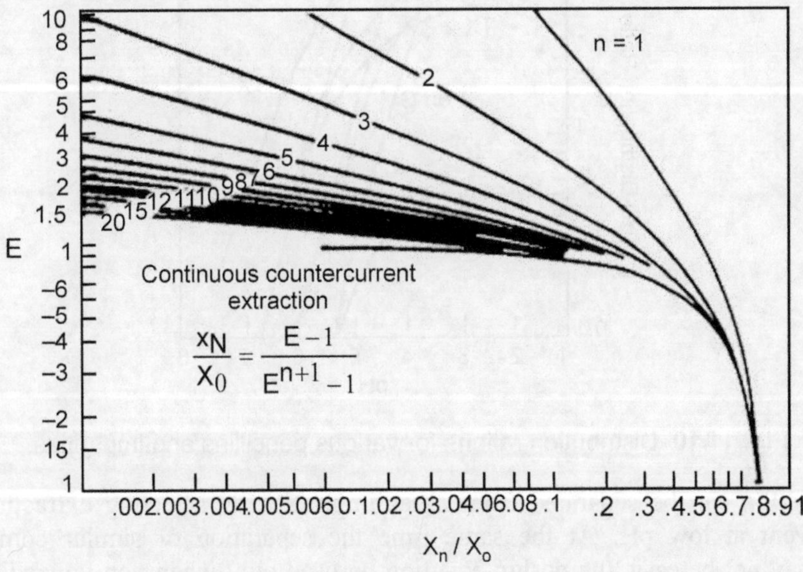

Fig. 11.9 Continuous counter current extraction – the relation ship between unextracted solute, extraction factor and number of stages.

Some products to be recovered from fermentation both could be weak acids or weak bases. The pH conditions are selected such that the extracted compound is neutral and is soluble in an organic solvents, since compounds are not ionized are soluble in the organic phase. So the weak bases are extracted at high pH values and weak acids are extracted at low pH values in the neutral form.

If more than one component is present in the fermentation broth, the selectivity (β) of the solvent would be an important factor. The selectivity coefficient (β_{ig}) can be defined as

$$\beta_{ig} = \frac{K_I}{K_f} \qquad\qquad .....(36)$$

β_{ig} value dictates the ease of the separation if the β_{ig} value is higher, the separation of 'i' will be easy from j. Some times β_{ig} values changes, appreciably with a pH shift and the separation will be easy Fig. 11.10.

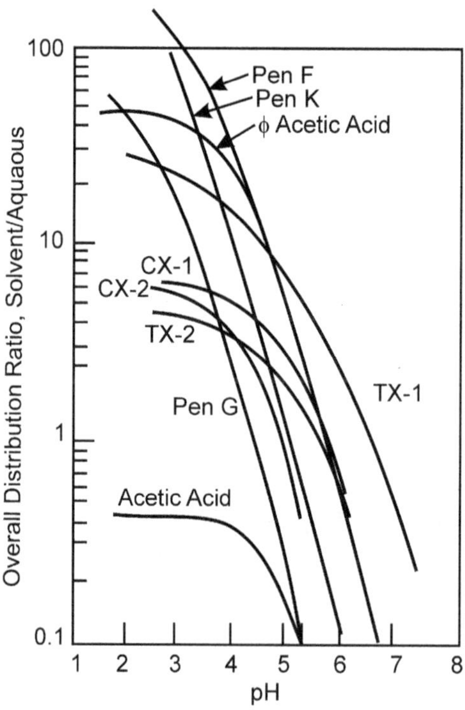

Fig. 11.10 Distribution rations for various pencillins and impurities.

The penicillin can be separated from other impurities by selecting extraction into the organic solvent at low pH. At the same time the separation of similar compounds by extraction may not be easy, the design equation be used only when non ionized species are soluble in both phases at extreme pH values. The extraction of ionized species i.e., weak acids or weak bases is known as dissociation extraction. The distribution coefficients per weak acids and bases may be given as.

For weak acids

$$K_D^{AP} = \frac{K_D^0 \left[H^+\right]}{\left[H^+\right] + K_1}$$

Or $$PH - PK_1 = \log\left(\frac{k0}{K_D^{AP}} - 1\right) \qquad\qquad .....(37)$$

For weak bases

$$K_D{}^{AP} = \frac{K_D^{\circ} K_1}{K_1 + [H^+]}$$

Or

$$PK_1 - PH = \log\left(\frac{K_D^0}{K_d}\right) - 1 \qquad \dots..(38)$$

Where

$K_D{}^{AP}$ = Apparent distribution coefficients

$K_D{}^{\circ}$ = Distribution coefficient for neutrals species

K_1 = Dissociation equilibrium constant per weak acids or bases

Generally liquid–liquid extraction for fermentation products is carried out by Podbielneak centrifugal extractions Fig. 11.11.

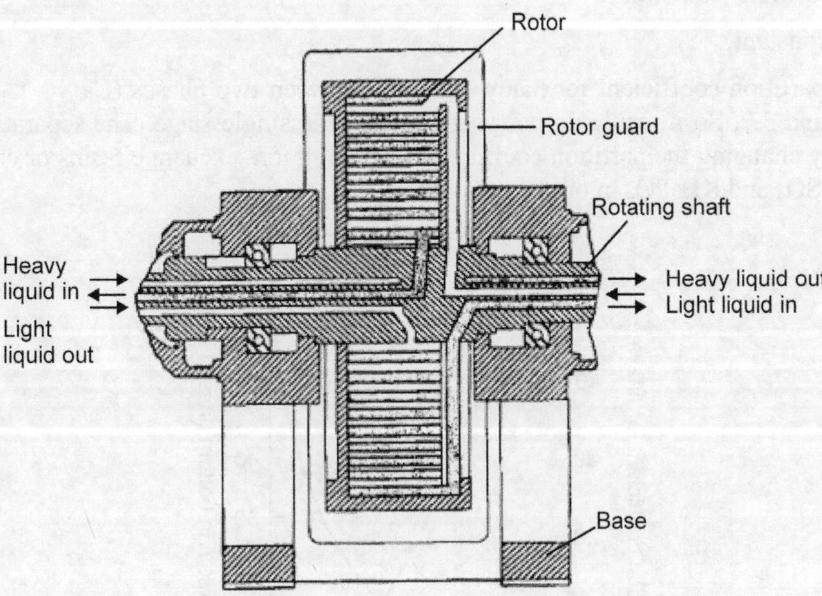

Fig. 11.11 Podbielneak centrifugal extractor – separator.

The Podbielneak centrifugal contactor provides two liquid introduction points through the rotating shaft. First, light liquid is forced to outside, the heavy liquid is introduced near the center. The rapid rotation facilities centrifugal force to move the fluids past each other in a counter current contacting mode.

Many fermentation products like pencillins are unstable. The rapid rotation of Podbielneak centrifugal extractor gives a centrifugal fluid that can rapidly drive the two fluids counter-current to each other as given Fig. 11.11. So a product can be extracted and sent to another aqueous phase i.e., phosphate buffer within a short time.

11.6 Two Phase Extraction

An aqueous two phase extraction may be the method for the extraction of soluble protein such as enzymes between two phases which have incompatible polymers such as Polyethylene glycol (PEG) and dextran. The phase that have PEG and dextran contain more than 75% water and they are not miscible. Aqueous phases generally used for this purpose are PEG – Water / dextran – water and PEG – Water / K- Phosphate – Water. PEG / dextran and PEG / K Phosphate are mostly immiscible. However the partition coefficient K_p, varies with the molecular weight of the soluble protein exponentially.

$$K_P = e^{AM/T} \qquad\qquad .....(39)$$

Where

M = molecular weight of the proteins

T = absolute temperature

A = constant

K_P, the partition coefficient for many enzymes between two phases ($C_{PEG} - C_{DEX}$) varies between 1 and 3.7, So it results in poor separations in a single stage. The separation can be improved by changing the partition coefficient by adding ion–exchange resins or certain salts like $(NH_4)_2\, SO_4$ and KH_2PO_4 in one of the phases.

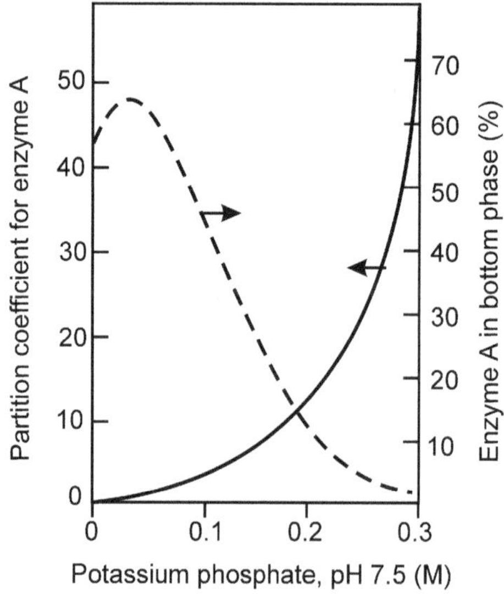

Fig. 11.12 The behavior of an enzyme in two phase solution.

Fig. 11.12, shows the variation of K_P (Partition coefficient) of an enzyme with a concentration of KH_2PO_4 in $PEG/(NH_4)_2SO_4$ system. K_P partition coefficient may be increased by (10) ten folds by increasing concentration KH_2PO_4 from 0.1 to 0.3 M in PEG – Salt systems, salting out may result in protein precipitation at the interface. K_P partition coefficient may be increased by addition of ion exchange resins.

K_P values may also be increased by adding affinity legends like PEG – NADH and PEG – Cibacron blue system. This method is known as two aqueous phase affinity partition extraction. However partial hydrolysis of dextran and PEG may increase K_P value, as lower – MW polymers may interact with proteins more efficiently. Also by mixing two kinds of PEGs (PEG_{400} and PEG_{4000}), the K_P value may increase by a factor of 6 for the partition of fumarase. It is interesting to note that the two – aqueous phase extraction may also be used for the recovery of cell derbies, Polysaccharides and nucleic acids. The partition coefficient for whole cells and DNA is between 100 and 0.01, for whole cells and partition coefficient for proteins it is between 10 and 0.1 and for small ions it is around 1.

Once the extraction is completed, the phases can be separated by centrifugation or decantation and PEG can be recovered by ultra filtration. In Fig. 11.13.

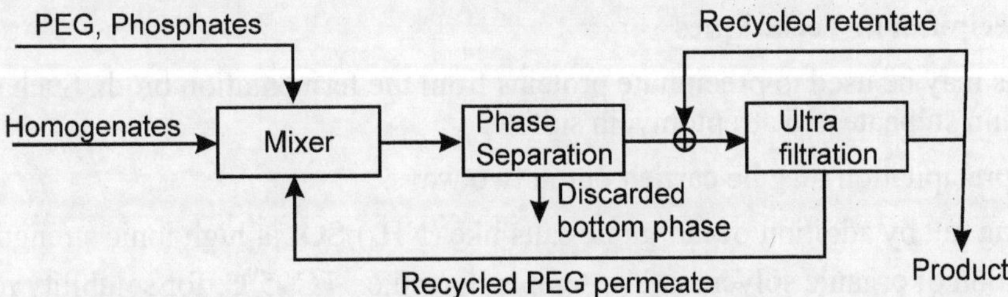

Fig. 11.13 Two phase extraction process with polyethylene glycol.

it is shown that the enzyme is separated by using two phase partition method and PEG recovery by ultra filtration. These separation process occur fast under mild temperature, pressure and pH conditions. PEG and dextran thus recovered are reused since the cost of PEG and dextran is high.

11.7 Precipitation

Precipitation may be carried out in various ways.

1. Addition of precipitant:- The solute reacts with the precipitant and insoluble product is formed mostly as a salt.

    Procaine – hydrochloride

$$\text{Sodium buffer salt} + \text{PENCILLIN} \rightarrow \begin{bmatrix} \text{Procaine} \\ \text{Sodium} \\ \text{Patassium} \end{bmatrix} - \text{Penceillin}$$

    Potassium buffer salt

Organic solvent + streptomycin + H_2SO_4 ------→ di – hydro – streptomycin sulfate.

Organic solvent + H_2O + erythromycin ------→ erythromycin hydrate.

Salt also may be added to recovery biopolymer. Xanthan gum is polyanion. Addition of divalent cation (calcium) may be carried out to form a gel precipitate. Alginate biopolymer is recoverable from algal biomass by cell removal (filtration) followed by a calcium chloride precipitation of the biopolymer.

Solvent Driven Precipitations

These type of precipitations are good in production of microbial bio-saccharide like Xanthan gum and dextran. The bio-gum fermentations are aerobic in nature and often proceduced highly viscous broth. After the production of Xanthan, the Xanthomonas cells are killed by final broth pasteurization when potassium chloride is added, the gum polysaccharide is precipitated by methanol or isopropanol that is added. The cell containing precipitate is then dewatered, washed and dried and it contains no viable cells. Addition of acetone or   alcohol precipitates dextran from the fermentation broth. Solvent recovery will make the process economical while precipitating bulk bio-polysaccharides.

Protein Precipitation Techniques

Certain salts may be used to precipitate proteins from the fermentation broth. Such salts could be ammonium sulphate and streptomycin sulfate.

Protein precipitation may be carried out in two ways.

1. Salting out by addition of inorganic salts like $(NH_4)_2SO_4$ at high ionic strength.

2. Addition of organic solvent at low temperatures i.e., TC -5^0C, for solubility reduction.

The ionic strength of protein containing solution can be increased by adding Na_2SO_4 and $(NH_4)_2SO_4$ which results in salting out. Protein molecules are precipitated by the interaction of the ions added with the water and the solubility of protein in solution with reflected to ionic strength of solution is given as

$$\log \frac{S}{S_0} = -K_s^1(I)$$

.....(40)

Where

S = solubility of protein in solution (g/l)

S_0 = solubility of protein where I = 0

I = ionic strength of solution

K_s^1 = salting out constant 'f' (temp, pH)

Now the ionic strength of a solution is given as

$$I = \tfrac{1}{2}\ \Sigma\ C_i\ Z_i^2$$

.....(41)

C_i = molar concentration of ionic species

Z_i = charge of ions (valence)

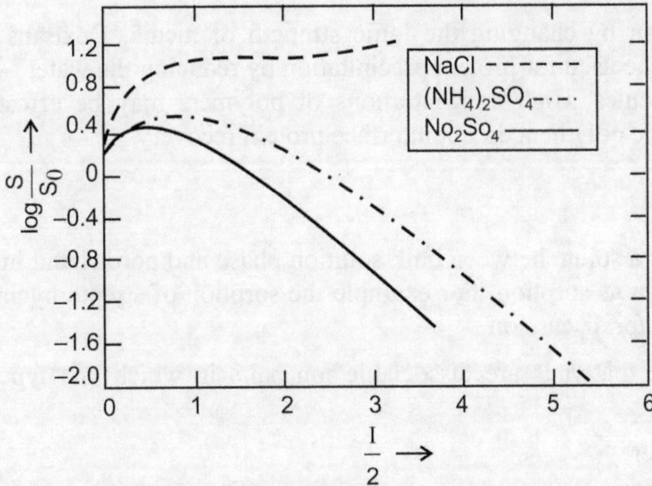

Fig. 11.14 Effect of inorganic salts on solubility of a typical protein. Fig (11.14) shows the solubility of hemoglobin with inorganic salt concentrations. The solubility of protein decreases logarithmically with increase ionic strength.

Precipitation of proteins by reducing the dielectric constant of the solution occurs as the addition of organic solvents at low temperatures i.e., T < -5⁰C. And the protein solubility as a function of dielectric constant of a solution is given as.

$$\log s/s_0 = - K^1/ D_s^2 \qquad \qquad \text{.....(42)}$$

where as

D_s = The dielectric constant of water – solvent solution protein

Precipitation takes place due to stronger electrostatic forces between protein molecules by the reduction of di-electric constant of a solution. Protein – water molecules inter actions may be reduced resulting in decrease in protein solubility due to the addition of solvents. But solvents may cause protein denaturation. In salting out method, protein denaturation may not take place. So the solvent precipitation may be used with the addition of salt, pH adjustment, and low temperature for improvement in precipitations. The other protein precipitation methods include, Isoelectric precipitation methods include, Isoelectric precipitation, the use of ionic polyectrolytes and use of non ionic polymers.

Isoelectric precipitation may be defined as the precipitation of protein at their isoelectric point which is pH at which proteins have no change. And isoelectric point of a protein can be given as

$$PI = 1/2 \, (PK_1 + PK_2)$$

Where PH = PI and protein becomes free of charge and precipitates.

This is effective when proteins have high surface hydrophobicity (non-polar surface) Though it is inexpensive, the low pH values may results in protein denaturation. Ionic polyelectrolytes like, polysaccharides, polyphosphates, and polycyclic acids may be used for protein precipitation by changing the ionic strength of media. Dextrans and polyethylene glycol (PEG) may be used for protein precipitation by reducing the water available to interact with protein molecules. High concentrations of polymers may be effective for low MW protein and nonionic polymers do not interface protein recovery.

11.8 Sorption

The partitioning of a solute between bulk solution phase and porous and high surface area of solid may be known as sorption. For example the sorption of streptomycin by ion exchange may be considered for discussion.

Ion – exchange materials are dissociable ion pairs in which one type of charge is not mobile.

Cation exchanger

$(solid)^-\ H^+ + solute$ ----------$\rightarrow$ $(solid)^-\ solute^+ + H^+$

Anion exchanger

$(solid)^+\ cl^- + solute$ -----$\rightarrow$ $(solid)^+\ solute^- + cl^-$

Streptomycin may be adsorbed in to a Carboxylic acid cation exchange resin

$$M\ (Resin\ )^-\ H^+ + streptorymycein^{\ n+} \text{------}\rightarrow\ n(\ Resin)^-\ streptomycin^{\ n+} + nH^+$$

The antibiotic or say streptomycin containing resin will be recovered from the fermentation broth, further the streptomycin is recovered by elution with acidulated water and regenerating the resin at the same time. Protein separations of small organics can be achieved by ion exchange resins and other polymeric adsorbents. The capacity of adsorption depends on adsorbent, adsorbate, physicochemical conditions and on the surface properties of adsorbent and adsorbate. The exact mechanism of adsorption is not clearly understood and experimental data at equilibrium may be determined.

The equilibrium relationship between solute concentrations in liquid and solid phases is given as

$$C^*_s = K_F\ C^{*\ (1/x)}_L \tag{43}$$

Where

C^*_L = equilibrium concentration of solute of solute in liquid

C^*_s = equilibrium concentration of solute of solute in solid

$$C_s = \frac{\text{Mass of soulte adsorbed}}{\text{Unit volume of resin}}$$

K_F, X = empirical volume

$X > 1$

The adsorption of solute may be carried at by various types of solid – liquid contactors. The example of solid liquid contactors are packed bed, fluidized bed, or agitated vessel contactors. The adsorption area per unit volume of reactor in packed and moving bed contactors is largest among other contactors, so they are used more frequently for various applications. Now adsorption of solute in a packed bed column based on the differential material balance of solution in column can be written as,

$$U\frac{\partial C_L}{\partial Z} + \varepsilon\frac{\partial C_L}{\partial t} = -(1-\varepsilon)\frac{\partial C_s}{\partial t} \qquad \qquad .....(44)$$

Where

μ = superficial velocity of the liquid

$\text{\textschwa E}$ = porosity of bed

C_L = concentration of the solute in the liquid phase

In equation (44) the convective transfer of solute in the bed in given by first term, and the time course of change of solute concentration in liquid is given by the second   term and the rate of the solute transfer from the liquid to the solid phase is given by the last term.

And the rate of solute transfer from the liquid into the solid phase is given by

$$\frac{dc_s}{dt} = K_a(C_L - C_L^*) \qquad \qquad .....(45)$$

Where

K_a = overall mass transfer coefficient (internal and external mass transfer resistance)

C_L^* = Concentration of solute in the liquid phase which is in equation with concentration of solute in solid phase C_s

From equation (45) and (44)

$$U\frac{\partial C_L}{\partial Z} + \varepsilon\frac{\partial C_L}{\partial t} = (1-\varepsilon)K_a(C_L - C_L^*) \qquad \qquad .....(46)$$

Equation (46) along with the equilibrium relationship

Equation (43) can be solved numerically to determine the solute profile $C_L = C_L(Z,t)$, in the column. The approach is based on plug flow conditions in the bed.

Generally due to back mixing and dispersion, irregular flow profiles and channeling takes place. A dispersion term $[-D_E(d^2c/dz^2)]$ is added to equation (46) to account for dispersion effects. So the fixed bed adsorption may be approximated to a large number of well mixed tanks in series.

For a very long moving bed columns to which absorbent added continuously, the operation may be considered to be at steady state as shown in Fig. 11.15.

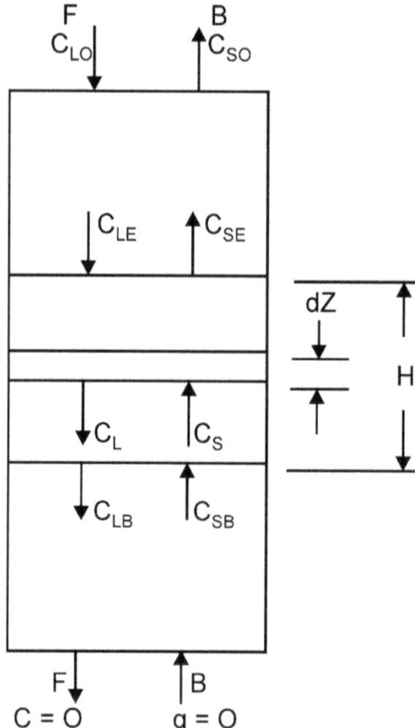

Fig .11.15 A flow of diagram steady state adsorption column.

The equilibrium is assumed to occur at the top of the column.

Now neglecting dispersion effects the equation (46) at steady state may be written as,

$$\mu\frac{\partial C_L}{\partial Z} = -K_a(1-\varepsilon)(C_L - C_L^*) \qquad \qquad .....(47)$$

The boundary conditions being,

$C_L = C_{Lo}$ at Z=H for liquid feed into the top of a column

So equation (47) can be integrated to determine the height of the column for a certain degree of separation,

$$\frac{-\mu}{K_a(1-\varepsilon)}\int_{C_L}^{C_{L0}}\frac{dc_L}{C_L - C_L^*} = \int_0^H dz = H \qquad \qquad .....(48)$$

Equation (48) can be integrated to determine the height of the column by using equilibrium relation ship. The analysis may be completed by drawing a relation ship between the unsaturated resin and the liquid concentration. And this relationship results an operating line which represents a mass balance Fig. 11.13 [scan Fig. 11.16 shuler pp353]

The volumetric resin flow is denoted by β and the volumetric liquid flow rate is denoted by F

So ,

$$F = U\,A \qquad \qquad .....(49)$$

Where

A = cross – sectional area of the column

Now from mass balance consideration on the solute we have,

Solute we have,

$$F\ (C_L - 0) = B\ (C^*_{s0} - 0) \qquad \qquad .....(50)$$

Or
$$\frac{F}{B} = \frac{C_L}{C_S} = \frac{C_{LO}}{C^{**}_{so}} \qquad \qquad .....(51)$$

Where

C^*_{so} = resin concentration of the solute in equilibrium with C_{L0}

11.9 Fixed Bed Columns

The absorptive capacity of fixed bed column is limited to maximum level. The column consists of three zones.

1. Saturation zones 2. Adsorption zones 3. a virgin zone

The feed rate of the solute and mass transfer characteristics of the bed determine the rate at which adsorption bed moves down the column.

11.10 Ultra Filtration

The solute particles such as proteins may be separated by means of membranes and membranes are used to separate used microbial cells from fermentation broth. Membranes are frequently used to separate proteins after precipitation from other contaminant particle. A variety of membrane can be used depending on molecular size cut off to separate protein of different molecular weights. Microorganisms like bacteria and yeast size in the range from 0.1 to 10 μm width, by using micro filtration or micro porous filtration. However macromolecules with a molecular weight range of 2000 to 500,000 are separated by using ultra filters. Ultrafilters may have anisotropic structure. A thin skin with small pores is formed on top, with a think highly porous structure in an anisotropic membrane. While the thick layer provides structures, the thin layer provides selectively. The membranes may be made of polymers or Ceramic materials. Most of the operations are basically pressure driven. The low – MW solutes and water pass through the membrane while high MW solutes are retained on the surface of the membrane. Eventually a concentration gradients builds up between the membrane surface and bulk fluid. The concentration gradients the developed results, in concentration polarization. So the solute diffuse back to the solution from the surface of the membrane Fig. 11.16.

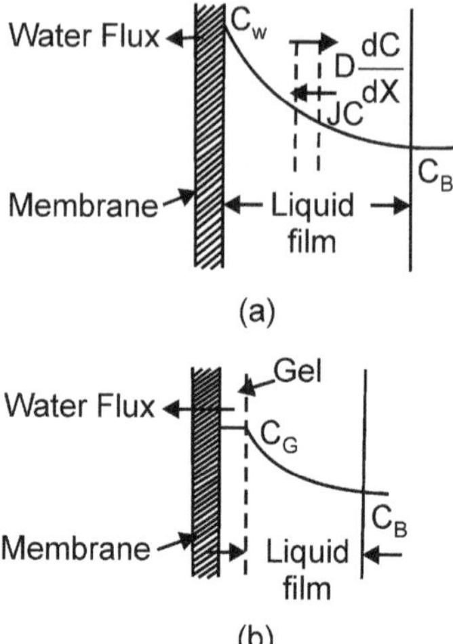

Fig. 11.16 Solute transfer in ultra-filtration membrane (a) with out gel formation (b) with gel formation.

And at steady state, the rate of convective transfer of solute towards membrane is equal to the rate of diffusion of solute in the opposite direction due to concentration polarization.

$$D_e \frac{dc}{dx} = J_c \qquad\qquad .....(52)$$

Where

    De = effective diffusivity of solute in the liquid film (cm^2/s)

    J = volumetric filtration flux of liquid (cm^3 / cm^2.s)

    C = concentration of solute (mol / cm^{3}liquid)

Upon the integration equ (52) at boundary conditions

    C = C$_B$ at   X = 0;

    C = C$_W$ at   X = δ;

We get,

$$J = \frac{D_e}{\delta} \ln \frac{C_W}{C_B} \qquad\qquad .....(53)$$

$$J = k \ln \frac{C_w}{C_B} \qquad\qquad .....(54)$$

Where

    K = D$_e$ / δ   mass transfer coefficient

    δ = film thickness

And the mass transfer coefficient is a function of fluid and solute properties and flow conditions. Mass transfer coefficient is calculated with Reynolds number (R_e) and Schmidt number (S_c)

$$(\text{Sherwood number}) \ \ Sh = \frac{dk}{D_e} = a \ Re^b \ Sc^b \qquad \qquad .....(55)$$

Where

$Re = dv\rho/\mu$

$Sc = \mu /\rho De$

$Sh = k_d/D_e$

And $a = Y_3$ according boundary layer theory

$b = 0.5$ for laminar flow

$b = 1.0$ for turbulent flow

So $k \ \alpha \ v^{0.5}$ for laminar flow

$k \ \alpha \ v$ for turbulent flow

when slowly diffusing large macromolecules accumulate at the surface ,a gel layer is formed on the surface of the UF membrane. Also when the protein concentration in the solution exceeds 0.1%, a gel layer is formed. Particularly for biological fluids, the protein accumulation is usually dominant. The protein concentration in the gel C_G is the maximum value of C_W

Here the liquid flux through the filter is given by

$$J = k \ln \frac{C_G}{C_B} \qquad \qquad .....(56)$$

Generally the gel formation depends on the nature and concentration of the solute, pH and pressure. Once gel is formed C_G becomes constant and liquid flux decreases logarithmically with increasing solute concentration in the bulk liquid. So the gel layer causes a hydraulic resistance against flow and may act like a second membrane.

The cross flow ultra filtration and micro filtration will not be discussed here.

11.11   Purification

Purification operations include fractional precipitation, many kinds of chromatography and adsorption. Out of the various operations mentioned, chromatography operation will be discussed here.

11.12   Chromatography

Mixtures are separated into components, in chromatography by passing a fluid mixture through a bed of adsorbent material. In elution chromatography a column is packed with adsorbent particles which may be solid, a porous solid, a gel or a liquid phase immobilized, in or on a solid. Then a mobile phase or fluid phase with a mixture containing solutes is injected. This pulse is followed by a solvent or eluent and the pulse enters as narrows concentrated peak but exits dispersed and diluted by additional solvent.

Different solutes present in the mixture interact differently with the absorbent material i.e., the stationary phase, as some interact weakly and some interact strongly. The solutes which interact weakly with the matrix, pass out of the column rapidly Fig (11.17) and those which interact strongly with the matrix exit slowly.

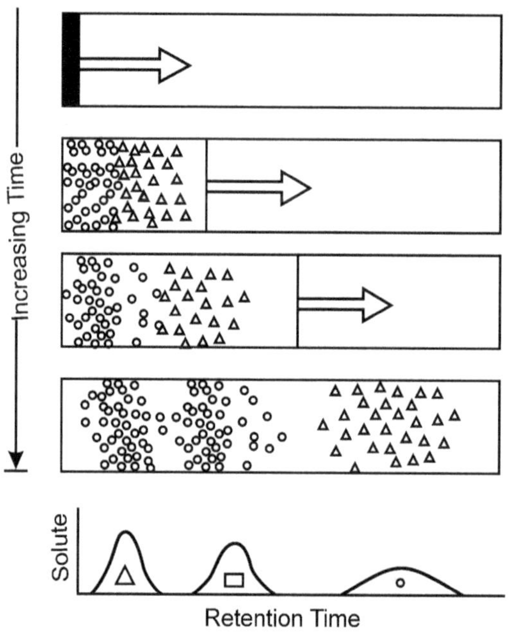

Fig. 11.17 Concentrations in elution chromatography.

Three solutes shown diagrammatically, as circles, squares. And triangles are injected into one end of a packed bed. When solvent flows through bed from left to right, the three solutes move at different rates due to different adsorption. They exist at different times, and they are separated.

An Agilent 6890 gas chromatograph equipped with a flame ionization detector, an Agilent 7683 series injector and auto-sampler

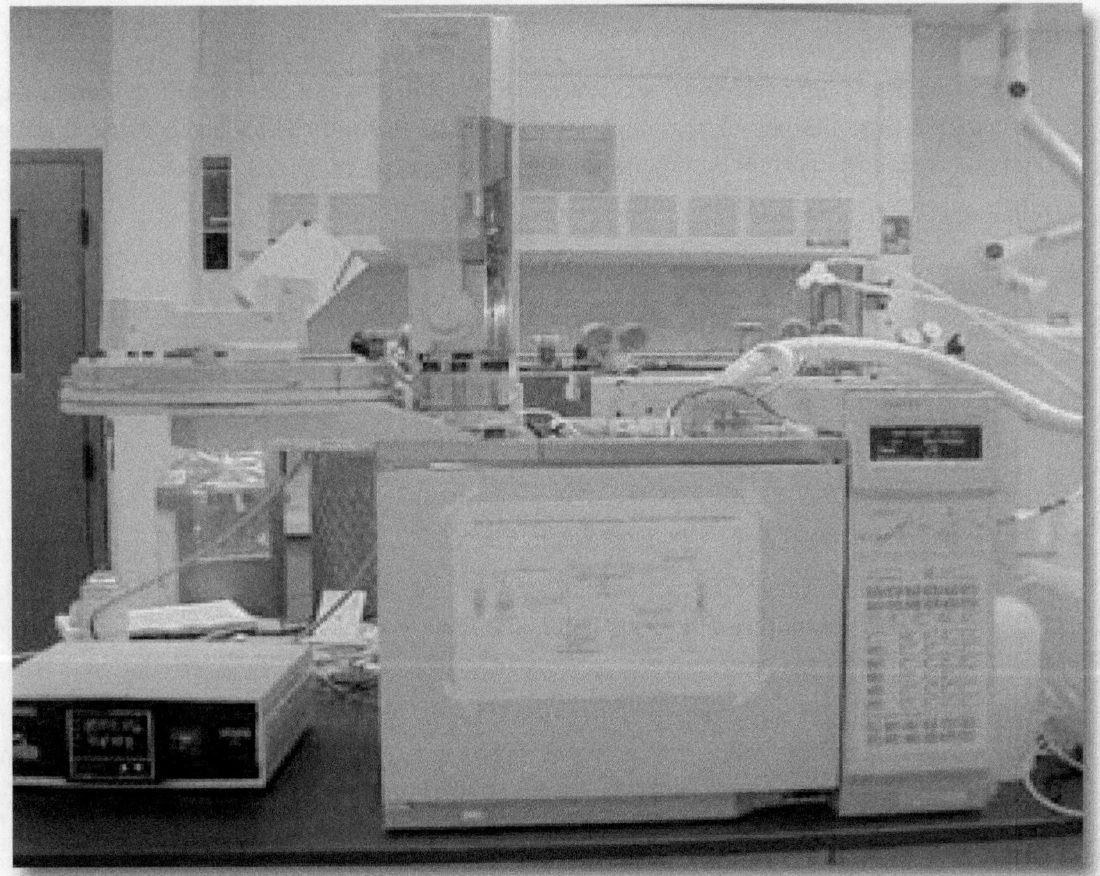

Fig. 11.18 Gas chromatograph.

The various peaks that are formed at a characteristic retention time, indicate certain solutes which migrate at differential rates. The similar type of separations are elution chromatography and fixed bed adsorption. In fixed bed adsorption the solute is captured in the adsorbent and then eluted as concentrate, while in elution chromatography product is purified as it is diluted.

Displacement chromatography will be useful while dealing with the large amounts of materials. This method involves the changes in inlet conditions step wise in sequence. The feed mixture is introduced prior to addition of displacer solution. The displacer is chosen such that it has more affinity towards the stationary phase, than any other compound in the feed solution. The solute is pushed off by the displacer from the stationary phase to place it in mobile phase. However, the feed components are forced into adjacent square wave like zones of the concentrated pure solutes when the conditions are maintained correctly. At the end of the column, these zones do break and the zone that has solute with the affinity for the stationary phase exists initially. The displacement chromatography is more advantages than the elution chromatography with respect to higher through put. However this operation is more difficult and high resolution is separation of solutes would be difficult in some cases. Most of the time the chromatography separations are carried out in batch mode, through continuous mode and semi-continuous made could also be tried.

Important types chromatography methods are

   (a)   Adsorption chromatography (ADC)
   (b)   Liquid chromatography (LLC)
   (c)   Ion exchange chromatography (IEC)
   (d)   Gel filtration or molecular sieving chromatography
   (e)   Affinity chromatography (AFC)
   (f)   Hydrophobic chromatography (HC)
   (g)   High pressure liquid chromatography (HPLC)

   (a)   *Adsorption Chromatography :* In this the solute molecules are absolved as the solid particles like, silica gel, alumina by weak vanderwaals forces and strict interactions.

   (b)   *Liquid Chromatography :* In this the solute molecules having different partition coefficient solution, between an absolved liquid phase and passing solution . Adsorbed liquid is non polar in nature.

   (c)   *Ion exchange chromatography :* In this the ions or electrically charged are adsorbed on the ion exchange resin by electrostatic forces.

   (d)   *Gel Filtration / Molecular sieving chromatography :* In this the solute molecules are penetrated into small pores of packing particulates depending on the molecular size and shape of the solute molecules.

   (e)   *Affinity Chromatography :* In this the separation takes phase depending on the specifies chemical interaction between molecules covalent like to a support particles.

   (f)   *Hydrophobic Chromatography :* In this the hydrophobic interaction takes place between solute molecules and functional groups on support particles.

   (g)   *High pressure liquid Chromatography :* In this high liquid pressure is applied to the packed column. So due to high pressure liquid and high liquid flow rate and dense column packing , HPLC gives fast and high resolution of solute molecules.

   For recovery of molecules like proteins, ion exchange chromatography is frequently used.

Affinity Chromatography

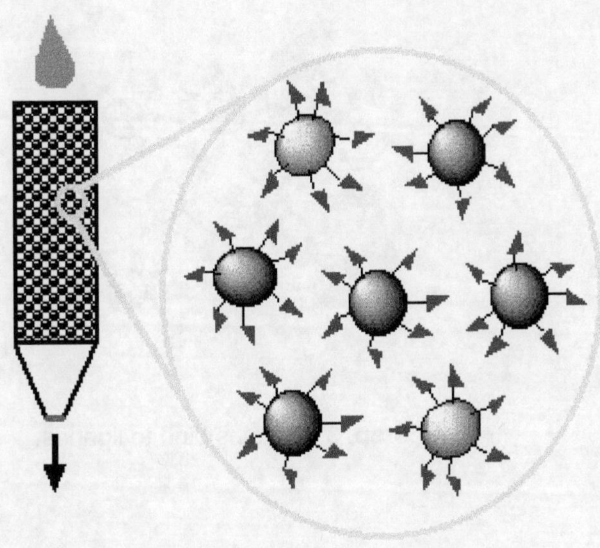

Step. 1 Attach ligand to column matrix.

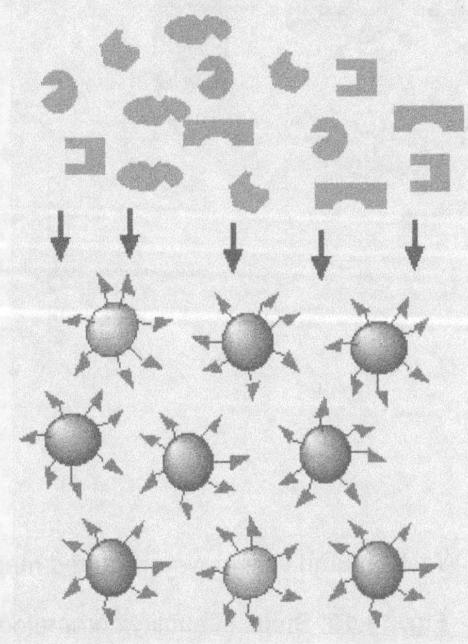

Step. 2 Load protein mixture onto column.

Affinity Chromatography

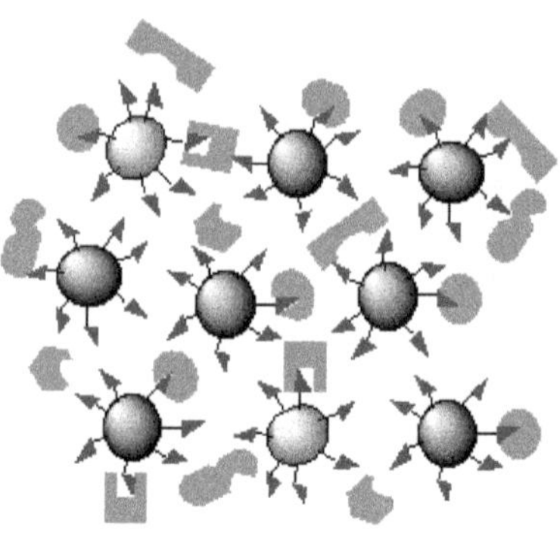

Step. 3 Proteins bind to ligands.

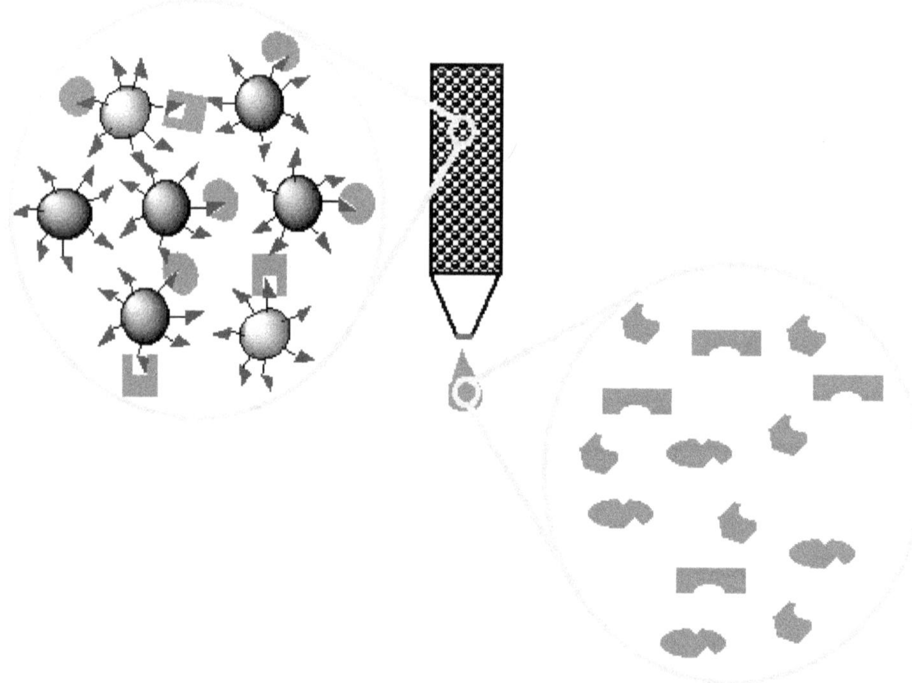

Step. 4 Wash column to remove unwanted material, elute later.

Fig. 11.19 Steps in affinity chromatography.

The following table (11.1) gives a brief description of various chromatography methods.

Table 11.1 Various chromatographic methods.

S. No.	Chromatography	Applicability
1.	Adsorption Chromatography	Relatively inexpensive how ever resolution is not sharp always
2.	Liquid Chromatography	These chromatography have similar applicabilities when aqueous phase is at high ionic strength, they are applied organic solvents may be required during process which may denaturate proteins or result in environmental process.
3.	Hydrophobic Chromatography	
4.	Ion exchange Chromatography	More frequently used like per recovery of proteins and this method has good resolution peaks, high capacity and good speed
5.	Gel filtration Chromatography	This method is good for buffer exchange and desalting and offers decent resolution .since gels are compressible, capacity and through put is always low.
6.	Affinity Chromatography	These methods offer high selectively good capacity and speed. If based on antibiotics these methods may be expensive.
7.	HPLC	Provides fast and high resolution of solute molecules

Often a chromatographic separation may include more then one of the mechanisms mentioned in the Table (11.1). For instance the adsorption chromatography, when water content of silica gel is less than 15%, functions well, but the system will operate as liquid liquid partition chromatography when the water content of silica gel is more than 30%.

A chromatography column is represented in Fig. 11.20.

In this the solute is adsorbed on adsorbent solids is by forming a band and solution flows down the column. Solute is carried by solvent when moving down would along the axis of the column. A rapid equilibrium between the solute and adsorbent is assumed while the radial gradients are considered negligible. A small volume of solvent ΔV, is poured through a band, and the solvent carries the adsorbed solute downward the column.

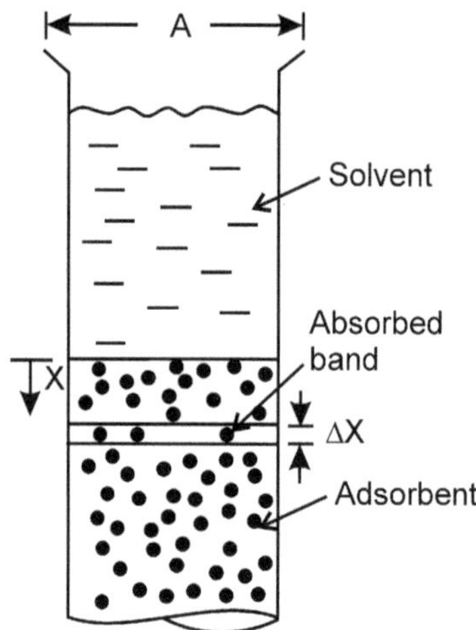

Fig 11.20 A diagrammatic representation of chromatography column.

A material balance on solute over a differential column height ΔX gives

Total rate of		Rate of solute		Rate of solute
Solute removal	$=$	removal in	$+$	removal in
From solvent		void space		in solid phase

$$-\left[\frac{\partial C_L}{\partial x}\Delta x\right]\Delta V = \varepsilon A\Delta x\left(\frac{\partial C_L}{\partial V}\right)\Delta V + A\Delta x\left(\frac{\partial C^1_s}{\partial V}\right)\Delta V \qquad .....(56)$$

Where C_s = amount of adsorbed solute per unit volume of the column

On simplification of equation (56) we have

$$\frac{\partial C_L}{\partial x} + A\left(\varepsilon\frac{\partial C_L}{\partial V} + \frac{\partial C^1_s}{\partial V}\right) = 0 \qquad .....(57)$$

The amount of solute adsorbed per unit volume of column is closely related to the adsorption isotherm of the solute

$$C^1_S = M_f(C_L) \qquad .....(58)$$

Where

M = amount of adsorbent per unit volume of column (mg/ml)

f (C_L) = adsorption isotherm or amount of adsorbed solute per unit amount of absorbent

Now by substituting eqn (58) in eqn (57)

We get

$$-\frac{\partial C_L}{\partial x} = A\left[\varepsilon + M\ f^1(C_L)\right]\frac{\partial C_L}{\partial V} \qquad .....(59)$$

Now upon substitution of

$$\frac{\partial C_L}{\partial x} = \left(\frac{\partial C_L}{\partial V}\right)\left(\frac{\partial V}{\partial x}\right)_c \quad \text{into equation (59)}$$

Gives,

$$\left(\frac{\partial V}{\partial x}\right)_c = A[\varepsilon + Mf^1(C_L)] \qquad \qquad .....(60)$$

Integration of equ (60)

From $X = X_0$ to $X = X$ and $V = V_0$ to $V = V$ gives

$$\Delta x = \frac{\Delta V}{A[\varepsilon + Mf^1(C_L)]} \qquad \qquad .....(61)$$

Now ΔV, A, ε and M are constants

ΔX is determined by $f'(C_L)$ which is derivative of $f(C_L)$ an adsorption isotherm.

So the location of the solute band formed is mainly determined by the adsorption characteristics of a particular solute in an adsorbent.

So when the solvent contains two components I and J, the solutes form distinct band at different axial distances depending on their characteristics of adsorption. Fig. 11.21.

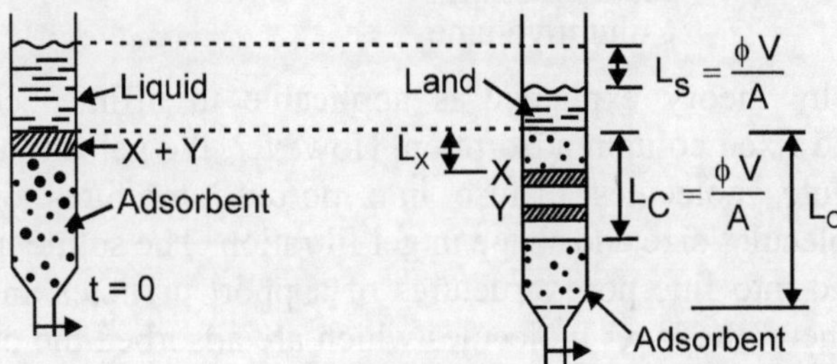

Fig. 11.21 The separation of binary mixture in a chromatographic column.

Now if L_I is the distance traveled by solute I, L_s is the distance traveled by solvent S alone the adsorption column $(\Delta V/A)$

L_C is the penetration distance of solvent into the column $(\Delta V /A \varepsilon)$

Then the resistance

$$R = \frac{\text{distance travelled by solute}}{\text{distance travelled by sovent in space above column}}$$

May be written as

$$\frac{L_I}{L_S} = R = \frac{1}{\varepsilon + MK} \qquad \qquad .....(62)$$

Where

$$K = f^{1}(C_L) = d\, f\,(\, C_L)\, /dC_L = \text{partition coefficient}$$

And R_f, may be defined as

$$R_f = \frac{L_I}{L_c} = \frac{\varepsilon}{\varepsilon + Mk} \qquad \qquad .....(63)$$

Where

$$R_f = \frac{\text{distance travelled by I in column}}{\text{distance travelled by S in column}}$$

Now when I is moved out of the column

$$L_I = L_0 \quad \text{and} \quad L_s = V_0/A$$

Where $V_0 = $ volume of solvent required to displace out of packed column

So L_s/L_I will be

$$V_r = \frac{L_s}{L_I} = \frac{V_0}{L_0 A} = \varepsilon + Mk \qquad \qquad .....(64)$$

Where

$$V_r = \frac{\text{solvent volume}}{\text{Column volume}}$$

The chromatography theory explained is applicable to affinity chromatography, Ion exchange, partition and fixed column adsorption. However, the gel filtration chromatography is different. The solute molecules diffuse into porous structures of support particles, depending on their molecules size and shape in gel filtration. The solute molecules of smaller size well be penetrated into fine pore structures of support particles and remain there with solid for long time, where the lager molecules which are adsorbed the outer surface remains there with solid for a short time. As the column is eluted the larger molecules comes fast is eluent solvent and smaller particles comes last in eluent solvent. The solute molecules of different bands are obtained depending on their shape and size. It may be condensed here the case where a buffer with solute is added to the column and elution of buffer solution due to the sufficient washes.

So the total volume buffer solution eluted from the column over a period of time will be

$$V_e = V_0 + K_D V_i \qquad \qquad .....(65)$$

Where

$\quad V_e = $ Volume of total element buffer

$\quad V_0 = $ void volume of column (V_ε)

$\quad V_i = $ total void volume inside gel particles

$\quad K_D = $ partition coefficient

Now for large molecules that do not penetrate inside gel structure

$$K_D = 0$$

And small molecules that completely penetrate inside gel have $K_D = 1$

Equ (65) can be written as

$$\frac{V_e}{V_0} = 1 + K_D (V_i/V_0) \qquad \qquad .....(66)$$

Now for a given solute mixture and gel beads

we may fix K_D and V_i/V_0 and

$V_e/V_0 = $ constant is irrespective of column geometry and V_e

The molecular weight of the macromolecules may be determined by gel filtration.

A plot of V_e/V_O Vs log (Mw) yields a straightly line with a negative slope which is proportional with K_D. The interaction which is highly specific between solute molecules and ligands attached on polymeric or ceramic beads in a packed column is used in affinity chromatography. The principles of affinity chromatography is shown in Fig. 11.22.

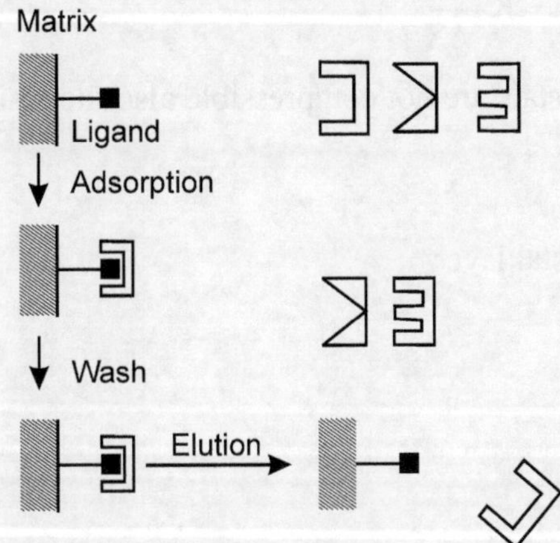

Fig. 11.22 Basic principles of affinity chromatographic separations.

Agarose may be used as matrix bead. However polyacrylamide, hydrooxy ethyle methacrylate, cellulose porous glass may be used as matrix. Matrix and ligands here linear aliphatic hydrocarbons arms. The stearic hindrances generated by the matrix may be reduced by use of space arms. The functional group present in the matrix and ligand dictated coupling between matrix and ligand. The functional group like –OH, -NH$_2$, or COOH groups are the chemical reactive groups are presented in matrix. Cyanogen bromide (CNBr) may be used as coupling agents for reactive groups like OH in matrix (poly saccharides etc). The cyanogens bromide activated activated agarose reacts with proteins that act as ligands.

The elution is achieved by changing pH or ionic strength in the column after the desired solutes are bound to the ligand. Ligand solute molecule interactions in affinity chromatography are specific like interacts of enzymes substrate or antegen – antibody of course an enzyme inhibitor or substrate may be used a ligand in separating a specific enzyme from mixture. All so monoclonal antibodies (MAB) may be used ligand to separate specific antigen molecules by affinity chromatography.

Final Product Isolation

The various purification steps of the fermentation product include crystallization and drying.

11.13   Crystallization

Most of the purified fermentation products like antibiotics are produced in crystalline form. Since crystallization of this type of products takes place at low temperatures, heat sensitive products are protected from thermal destabilization. These operations are carried out at high concentrations so the unit costs are low and have have separation factors. Either Nutsche – type filters or centrifugal filters are used for recovering highly pure crystals.

The following equation may be used for the filtration of crystals

$$\frac{dv}{dt} = K\frac{A\Delta P}{\Delta X} \qquad .....(67)$$

It is assumed that the crystals are not compressible also the resistance of filter medium is negligible.

Where

ΔX = thickness of crystal layer

t = time

V = liquid volume

ΔP = pressure drop

K = transfer coefficient

Substituting $W\ \alpha\ V$

W = solids concentration in cake

Upon integration equation (67)

$$t = K\frac{w^2}{\Delta P A^2} \qquad .....(68)$$

The filtration, crystallization the rate of centrifugal filter may be given by the following equation

$$\frac{dv}{dt} = \frac{(2\Pi N)^2 \rho L(r_0^2 - r_L^2)}{2\mu(\alpha w / A_m^2 + R_m / A_0)} \qquad .....(69)$$

Where

N = rotational speed

α = specific cake resistance

w = solids concentration

A_m = logmean Area

R_m = medium resistance

The crystal size and nature affect the centrifugation and washing rates. The crystals are sent for drying after the washing operation.

11.14   Drying

During drying process wet crystals is dried by removal of solvent. The desirable moisture content and heat sensitivity of product are considered for selecting a drying operation. The parameters that effect drying can be categorized into four groups.

1. Physical properties of solid – liquid system

2. Intrinsic properties of the solute

3. Conditions of drying environment

4. Heat transfer parameters

The dryers usually adopted per fermentation product are given in the following Table. 11.2.

Table 11.2 Various types of drying.

S.No	Dryer	Description
1.	Vacuum tray drier	It contains heated shelves in single chamber. This is used in Pharma industry and heat damage and product loss in minimized .
2.	Freeze drying (lyophilization)	In this water is removed by sublimation from frozen solution ie solid into vapor . Prier to drying, freezing may be adopted to a vacuum chamber. This method is used for bacterial suspension, enzymes solutions and antibiotics.
3.	Rotary vacuum dryer	This method is not good for solutions containing crystals. Water is removed by steam heated surface of rotary drum from the thin fraction of the solution. The dried product is scrapped by a knife
4.	Spray dryer	These are used for heat sensitive materials product solution is sprayed through a nozzle with heated chamber .
5.	Pneumatic conveyer driers	These are used for heat substance and oxidizable materials, The retention time of a particle in gas steam is short. This method is good when surface drying is critical.

Fig. 11.23 Lyophilizer.

Fig. 11.24 Vacuum Tray dryer.

References

1. Aiba S., A.E..Humphrey and N.F. Mills (1973), Biochemical Engineering 2 Edn, Academic Press New York.

2. Bailey J.E and D.F Ollis (1986), Biochemical engineering Fundamentals, 2 nd ed, McGraw Hill N.Y.

3. Blanch, H.W and D.S. Clark (1996), Biochemical engineering Marcel Dekker, N.Y.

4. Kundu, A.K., A Barnthouse, and S.M. Crainer (1997), "Selective Displacement chromatography of Proteins" Bio eng.

5. Wang D.I.C etal (1979), "Fermentation and enzymes technology, John Wiley and Sons N.Y.

Review Questions

1. Write about prior crystals of separation.

2. Describe removal of Particulates.

3. Explain design considerations for centrifuge.

4. Describe filtration.

5. Write about sedimentation.

6. Describe disruption of cells by mechanical means.

7. Explain non mechanical methods of cell disruption.

8. Describe extraction.

9. Explain Precipitation.

10. Write about protein precipitate techniques.

11. Describe Sorption.

12. Explain ultra filtration.

13. Write about Chromatography.

14. Explain crystallization.

15. Describe drying.

CHAPTER 12

Scale up Operation

Scale up is desired whenever the laboratory studies are to be unchained to translated scale. Various equipments required in bioprocess industries like bioreactors, centrifuges, crystallizers, etc can be designed if scale up concepts are well understood. The scale up of bioreactors or fermentors of related to bioprocess are discussed here. Now consider the aeration in a bioprocess at laboratory scale is to be scaled up using geometrical similarities. It may be assumed that aeration in large bioreactors will be carried out without agitation.

$$\frac{(K_{La})_1}{(K_{La})_2} = \frac{(F/V)_1}{(F/V)_2}\frac{(H_{L_1})}{(H_{L_2})}\left(\frac{d_{B2}}{d_{B1}}\right)^{3/2}\left(\frac{v_{B2}}{v_{B1}}\right)^{1/2} \qquad \dots (1)$$

Where

Subscript 1 relate to small scale equipment

2 relate to large scale equipment.

K_{La} = volumetric oxygen transfer coefficient

F = aeration rate m^3/sec

V = liquid volume, m^3

H_L = liquid depth, m

D_B = bubble size, m

V_B = nominal superficial velocity of air m/hr

Since size of the bubbles may not differ with bioreactor size, it may assumed that $d_{B2} = d_{B1}$ and $V_{B2} = V_{B1}$. Aeration rate in the bioreactor may be found out by equating left side of equation (1) to be unity,

$$\left(\frac{(\frac{F}{V})_1}{(\frac{F}{V})_2}\right) = \frac{H_{L2}}{H_{L1}} \qquad \dots (2)$$

For example, the aeration rate $(F/V)_2$ with large bioreactor would be $(F/V)_2 = 5^{-1} = 0.2$ vvm, provided $(F/V)_1 = 1.0$ vvm and scale up ratio $H_{L2}/H_{L1} = 5$. The effect of oxygen transfer from the bubbling liquid surface and that associated with the transient state in the bubble's emergence at a sparger cannot necessarily be disregarded when bench scale bioreactor is used. Now to consider this effect it is assumed that the term (H_{L1}/H_{L2}) of equation (2) will be modified by $(H_{L1}/H_{L2})^{2/3}$. Considering this the aeration rate $(F/V)_2$ with large bioreactor would be $(F/V)_2 = 5^{2/3} = 0.34$ vvm instead of $(F/V)_2 = 0.2$ vvm.

During scale up of bioreactor, behavior of organisms in the reactor and metabolic reactions taking place inside the reactor are also considered.

12.1 Physical Concept

It is assumed that the physical properties like, temperature, pH, medium components, dissolved oxygen concentration would be similar between beach scale bioreactor and full scale bioreactor. Also it may be assumed that the microorganisms are will dispersed in turbulent zone in bench as well as full scale bioreactors.

Now consider agitated bioreactors for turbulent liquid,

$$P \propto n^3 D_I^5 \qquad \qquad \text{.....(3)}$$

$$P_g/P = f(N_a)$$

$$N_a = \text{aeration number } (= F/nD_I^3)$$

Where

P = power requirement of agitation

P_g = Power requirement of agitation in gassed systems

n = rotation speed of the impeller

F = aeration rate m3/sec

Pumping rate of liquid by impeller, m^3/sec

Also

$$V \propto D_i^3 \qquad \qquad \text{.....(4)}$$

$$F \propto n D_i^3 \qquad \qquad \text{.....(5)}$$

The physical terms that are considered for scale up are given here,

1. Power consumed per unit volume of liquid P/V

$$P/V \propto n^3 D_i^2 \qquad \qquad \text{.....(6)}$$

2. liquid circulation rate inside the vessel F/V

$$F/V \; \alpha \; n \qquad\qquad .....(7)$$

3. Impeller tip velocity v

$$\upsilon \; \alpha \; nD_i \qquad\qquad .....(8)$$

4. Modified Reynold number $nD^2_i \rho/\mu$

Where

$$\rho \; = \text{liquid density}$$

$$\mu = \text{liquid viscosity}$$

$$nD^2_i \rho/\mu \; \alpha \; nD^2_i \qquad\qquad .....(9)$$

from equation (6) we have,

$$n_2^3 D^2_{i2} = n^3_1 D^2_{i1} \qquad\qquad .....(A)$$

so $\qquad\qquad n_2/n_1 = (D_{i1}/D_{i2})^{2/3} = 5^{-2/3} = 0.34$

from equation (3) we have

$$P_2/P_1 = (D_{i2}/D_{i1})^3 = 5^3 = 125$$

For scale up ratio $D_{i2}/D_{i1} = 5$

For nD_i

$$n_2 D_{i2}/n_1 D_{i1} = 0.34 \times 5 \; = 1.7$$

Now considering equal power her unit volume for scale up, it is known that as the tip velocity of impeller increases the value of F/V or shear rate of liquid decreases. So as F/V decrease the mixing time increases.

If the scale up is carried out by considering equal power per unit volume eventually fermentation would be less sensitive to increase in tip velocity of impeller as well as mixing time.

The other factors include like heat transfer coefficient h, of liquid, with a coil in an agitator reactor, factors like, h, P/V and F/V will be of significance. And in some cases K_L mass transfer (oxygen transfer coefficient) will be of interest.

Now if the Sherwood number, $k_L D_i/D$ is replaced by the Nussselt number hD_i/k (h = heat transfer coefficient film; k = thermal conductivity) and Schmidt number $\mu/\rho D$ is replaced by Prandtl number $c_p \mu/k$ (c_p = specific heat) the analogous equation obtained can be used to calculate the liquid mixing performance in terms of heat transfer.

We consider the aeration condition in bioreactors such that aeration number Na is same or not different significantly from one system to the other which are geometrically similar. The power requirements of an impeller in gassed systems, can be represented in geometrically similar systems by P. (P = Power number in ungassed systems) see Fig. 12.1.

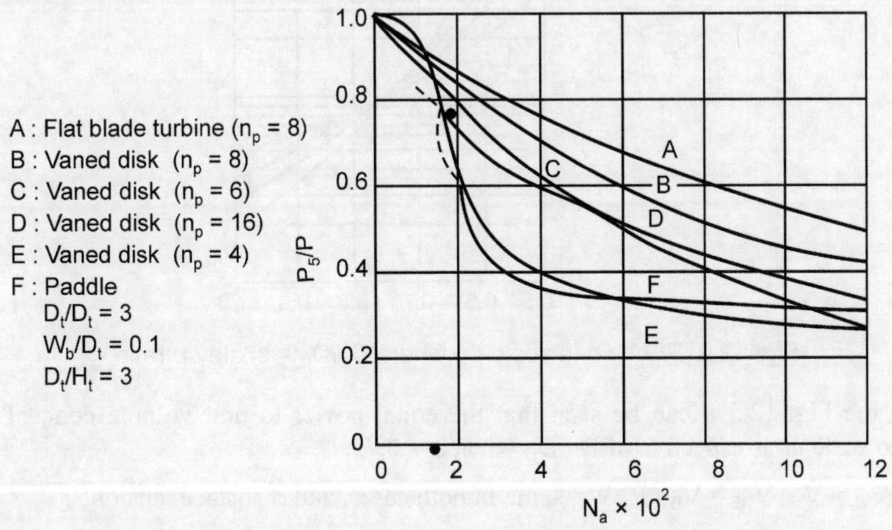

Fig. 12.1 Power requirements for agitation in gassed system.

P_G/P = degree of power decrease;

N_a = aeration number.

Assuming the physical properties of the liquid are constant, the K_L may be estimated by the following equation.

$$n_1/n_2 = (D_{i2}/D_{i1})^{(2\alpha-1)/\alpha} \qquad\qquad .....(10)$$

Now from equation (6)

$$\frac{P_2/V_2}{P_1/V_1} = (n_2/n_1)^3 \, (D_{i2}/D_{i1})^2 \qquad\qquad .....(11)$$

Upon cancelling n_1/n_2 from equation (10) and (11) we have

$$\frac{P_2/V_2}{P_1/V_1} = (D_{i2}/D_{i1})^{2-3\frac{2\alpha-1}{\alpha}} \qquad\qquad .....(12)$$

The left hand side of equation (12) is plotted against α, and the scale up ratio being D_{i2}/D_{i1} as shown in Fig. 12.2.

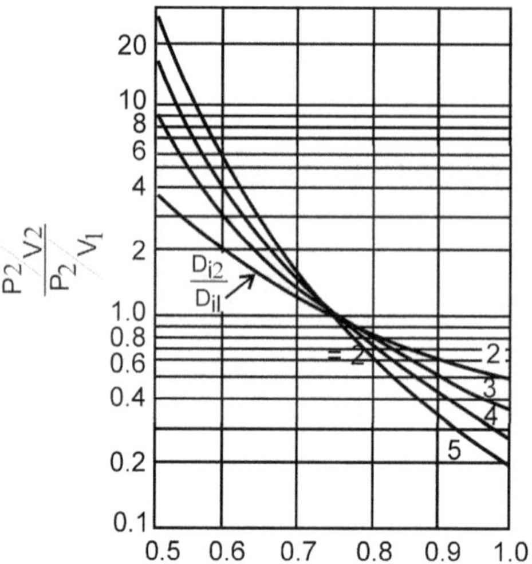

Fig. 12.2 P/V vs α in scale up where D_{i2}/D_{i1} = scale up ratio.

From the Fig. 12.2 it can be seen that the equal power to unit volume concept can be applied to scale up irrespective of D_{i2}/D_{i1} when $\alpha = 0.75$.

So if $V_{s2} = V_{s1}$, $V_{B2} = V_{B1}$, P_g/V = same in both cases, and σ surface tension,

ρ = density are constant, then the interfacial are a, between air bubbles and liquid per unit volume of liquid will remain unaffected in two bioreactors.

Particularly in microbial reactions or fermentations power consumed per unit volume P/V volumetric oxygen transfer coefficient K_{La} and the mean liquid velocity at a particular point in the vessel are used per scale up commonly. $P_g/p = f(N_a)$ is the criteria incorporated into the aeration number.

Now from,

$$a \, \alpha \left[\frac{(p_g/V)^{0.4}\rho^{0.2}}{\sigma^{0.6}} \right] (v_s/v_B)^{0.5} \text{ and}$$

$$K_{La}\alpha(P_g/v)^{0.4} v_s^{0.5} n^{0.5}$$

It can be inferred that the implications of using P/V (or P_g/V) and K_{La} as criteria of scale up of bioreactors do not differ appreciably as both criteria are some how connected with the rate of oxygen transfer from (directly/ indirectly) bubbles into the liquid. However P/V or P_g/V is closely related to air-liquid interfacial area a, per unit volume of liquid if α in equation $K_L D_i/D\alpha \, (nD_i^2\rho/\mu)^{\alpha} \, (\mu/\rho D)^{\alpha}$ deviates from 0.75 see Fig (12.2) for 'a' in equation

$$a\alpha \left[\frac{(P_g/V)^{0.4}\rho^{0.2}}{\sigma^{0.6}} \right] (V_s/V_B)^{0.5} \text{ and}$$

$$a\alpha(P_g/V)^{0.4} V_s^{0.5}$$

12.2 Biological Concept

The final product concentration is a bioreaction may be effected either by power input per unit volume of broth P/V by k_{La} volumetric oxygen transfer coefficient. The nature of the curve may be hyperbolic irrespective of the microorganisms like yeast, fungi, or bacteria (vide Fig. 12.3.)

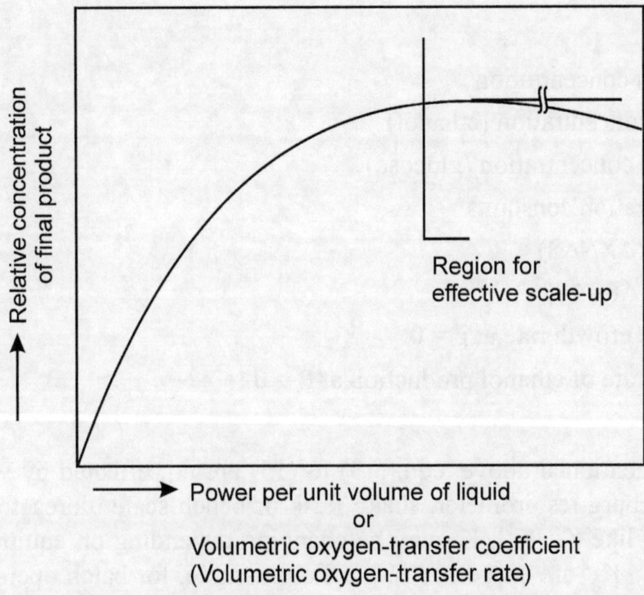

Fig. 12.3 Microbial performance and operation variables.

The optimum product concentration may be selected based on the various factors like, Power to unit volume or oxygen transfer coefficient; energy requirements or convenience of operation etc. Some times the other factors like dissolved oxygen concentration may also effect the yield of product, during scale up. The broken part of the curve may indicate effect of inhibition concentration or the final product formation. Also the effect of microbial metabolism on final product formation during scale up, has to be studied for various biological processes. This concept may be studied by considering ethanol fermentation.

Now considering continuous culture techniques like Chemostat pertaining to anaerobic culture, the kinetic equation may be given as,

$$\frac{dX}{dt} = \frac{\mu}{1 + \dfrac{p}{K_p}} \cdot \frac{S}{K_S + S} X \qquad \qquad .....(13)$$

$$\frac{dp}{dt} = \frac{v_0}{1+\dfrac{p}{K_p^1}} \cdot \frac{S}{K_s^1 + S} X \qquad \qquad \dots(14)$$

$$-\frac{dS}{dt} = \frac{1}{Y_{x/s}} \frac{dX}{dt} = \frac{1}{Y_{p/s}} \frac{dP}{dt} \qquad \qquad \dots(15)$$

Where

X = cell mass concentration

P = Product concentration (ethanol)

S = Substrate concentration (glucose)

K_s, K_s^1 = saturation constants

$Y_{X/S}$ = yield $(\Delta X/-\Delta S)$

$Y_{P/S}$ = yield $(\Delta P/-\Delta S)$

M_0 = specific growth rate at $p = 0$

μ_0 = specific rate of ethanol production at $P = 0$

T = time

The equations mentioned above, eqn. (13) to (15) are not effected by various patterns of culture like by Warburg respirometer, shake flask or bench scale bioreactors. Of course, the empirical constants like K_P, K_p^1, K_s may be changing depending on equipment during scale up. The value of K_p, K_p^1 are expressed in g/l. For example, for batch operation K_p for shake flask be 16 g/lit whereas for continuous reactor would be 55g/l. Similarly for batch Warburg respirometer K_p^1 is 71.5 g/l and for continuous process, it may be 12.5 g/l. So the specific rate of ethanol production and microbial activity of the yeast cells depends on the type of bioreactor.

So it is essential to study the microbial metabolism for fixing scale up parameters. And the empirical constants may be evaluated by experiments for any biological process and their relation with the operation variables like P/V or K_{La} etc in scale up.

12.3 Power to Unit Volume

The relation between product concentration and power input are shown in Fig. 12.4 (a) and (b).

The penicillin formed, is platted against power input (HP/m^3) is shown Fig. 12.4a and the 1:10 volumetric ratio of the scale up is considered. Also v_s is doubled approximately in each bioreactor. For the particular case presented at power input exceeding 1.5 HP/m^3 shown good results, during scale up while at power input less than 1.0 Hp/m^3 the yield of penicillin is reduced.

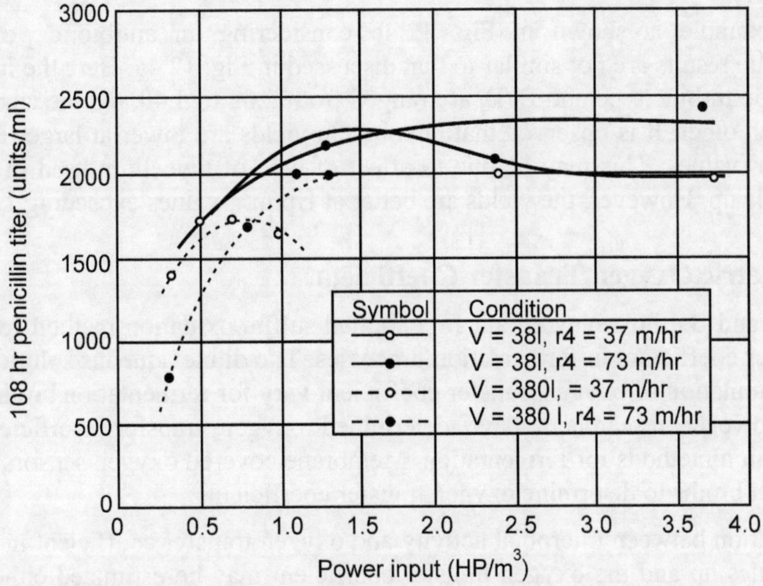

Fig. 12.4 (a) Effect of different values of power input on the yield of penicillin with different fermentation conditions.

If the power consumed per impeller in unit volume of liquid agitated taken as basis, the impeller performance may be evaluated accurately irrespective of geometric similarities during scale up.

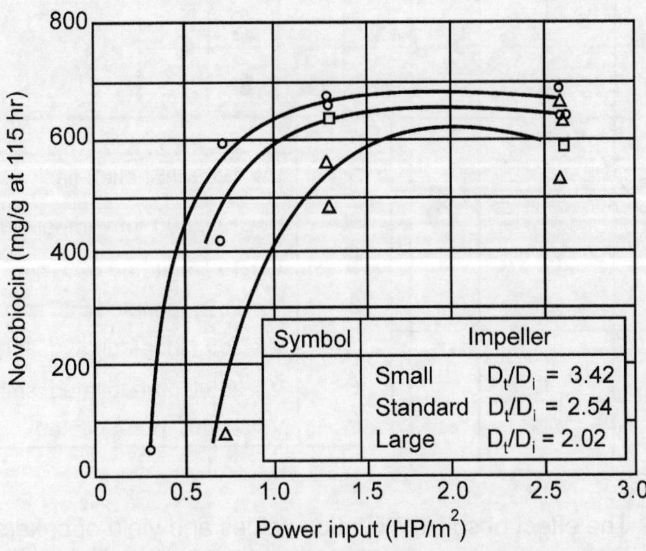

Fig. 12.4 (b) Effect of different values of power input on the yield of novobiocin in fermentors fitted with different size impellers.

Another example is shown in Fig. 12.4b considering an antibiotic production like novobiocin. The results are not similar to that discussed in Fig. 12.4a. Here the impellers used are of flat blade turbine type and D_t/D_i are ranged from 2.02 to 3.40 and the curves down are parallel to each other. It is observed that novobiocin yields are lower at larger impeller even at equal Hp/m^3 values. This may be due to effect of type of mycelium used in fermentation during the scale up. However, the yields are better at Hp/m^3., values exceeding 1.5 Hp/m^3.

12.4 Volumetric Oxygen Transfer Coefficient

Hixson et. al, and Bartholomew et. al., first applied sulfite oxidation method for finding out oxygen transfer coefficient in fermentation processes. The dilute aqueous solution of sulphite used for determination of oxygen transfer coefficient vary for fermentation broths in physical properties. However researchers have determined oxygen transfer coefficient by using sulphite oxidation methods for fermentation. Membrane covered oxygen sensors may be used in fermentation broths to determine oxygen transfer coefficients.

The correlation between microbial activity and oxygen transfer coefficient may be studied in view of scales up and the oxygen transfer coefficient may be estimated either or relative basis by using empirical equations available in literature. Some examples representing oxygen transfer coefficient and their effect on microbial product will be discussed in brief here. The yield of bakers yeast at different sulphite oxidation values are given in Fig.12.5.

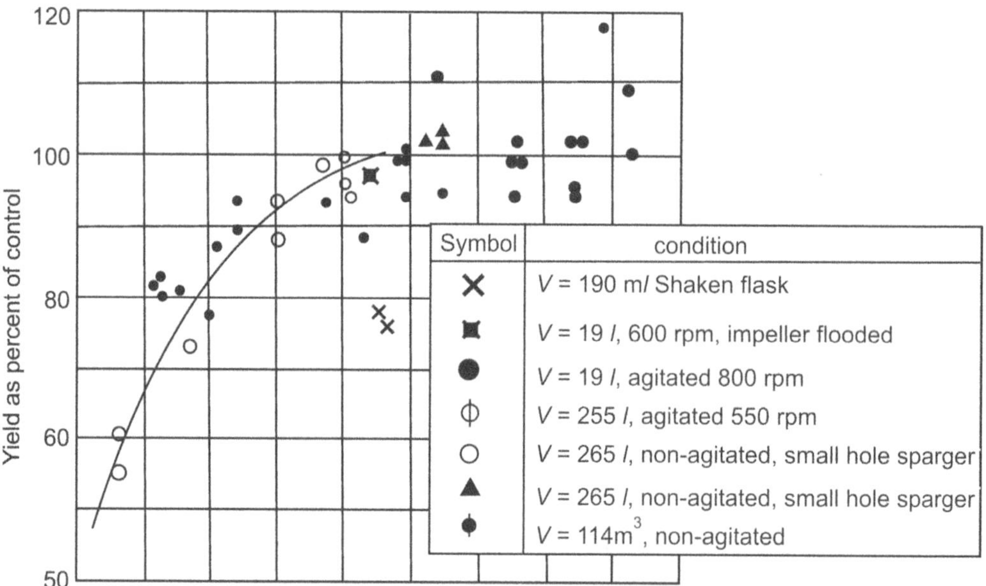

Fig. 12.5 The effect of sulfite oxidation values and yield of bakers yeast.

Even though the points are scattered in the Figure, the yield is correlated to the sulphite oxidation value to great extent. At sulfite oxidation values greater than 150 m mole O_2/l/hr, bakers yeast yield did not depend either on type of fermentation or method of aeration.

Another example is the production of ustilagic acid where product yield is correlated to sulfite oxidation value with reasonable approximation as shown in Fig.12.6.

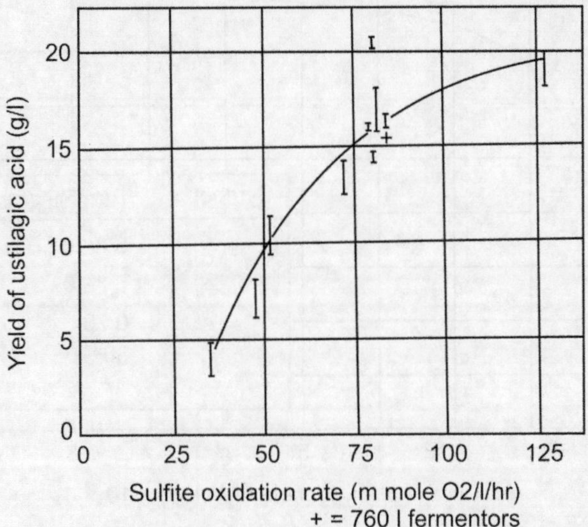

Fig. 12.6 The effect of sulfite oxidation value on the yield of ustilagic acid.

The experimental values plotted in Figure correspond to 5 *l* fermentation except with + symbol which corresponding to 760 *l* fermentor. The ustilagic acid yields with different degree of aeration and agitation for a medium contain 7.5% glucose mono hydrate are shown in Fig. 12.6.

The yield increased with sulfite oxidation value till it approached 125 mmole/ O_2/l/hr and above the values there is no improvement in yields which is similar to bakes yeast yield shown in Fig. 12.5.

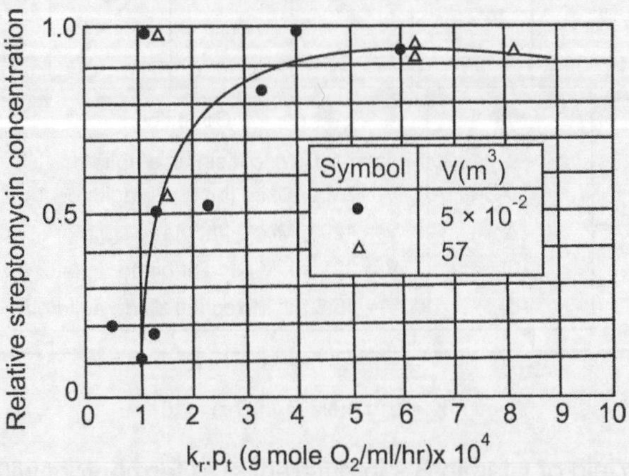

Fig. 12.7 Comparison of yields of streptomycin at different mass transfer coefficients of applying in vessels of different size.

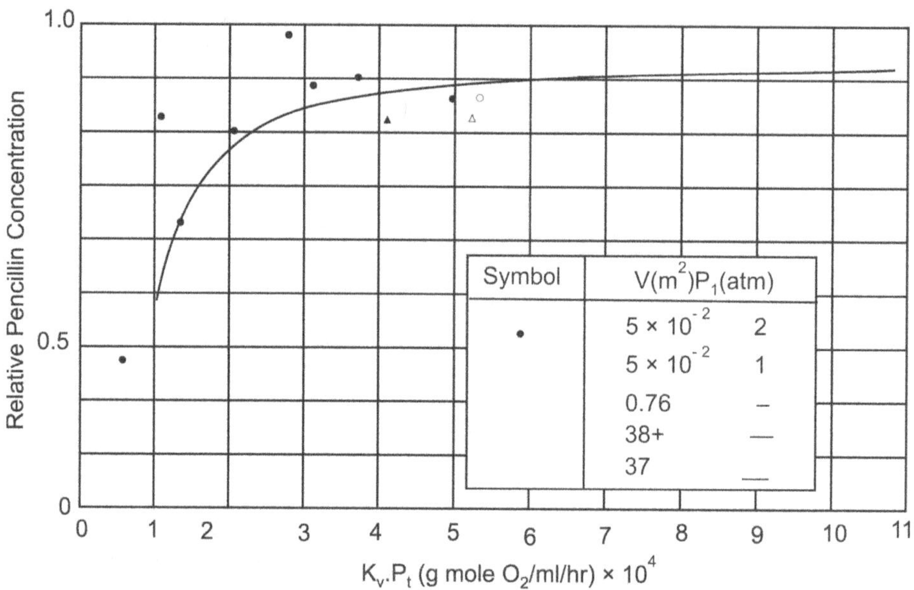

Fig. 12.8 Comparison of yields of penicillin at different mass transfer coefficients of applying in vessels of different size.

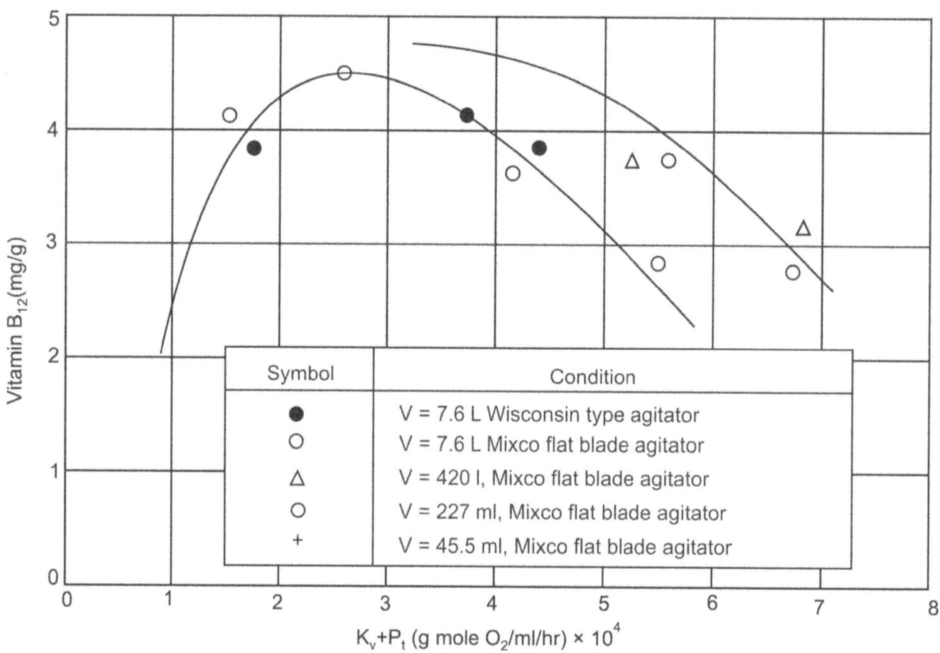

Fig. 12.9 Yield of vitamin B_{12} in different size fermentors with different values of mass transfer coefficients.

In the figures 12.7 to 12.9 the value of the Kv pt (volumetric oxygen transfer coefficient total pressure) are shown in abscissa. The pressure factor is considered in these figures. Since pressure increases greatly in large scale bioreactors, the K_v value may be determined by using empirical equations. The product K_v and pt evaluated by empirical values correlate with experimental values for streptomycin and penicillin. The scale up of vit B_{12} fermentation in Fig. 12.9.

Even though various models are used for translating laboratory studies to commercial scale by using scale up concepts, it has become essential to do vice versa by using scale down approach. Many parameters can be tested more quickly and inexpensively at smaller scale than at the commercial scale. These small scale studies will be useful in studying proposed process changes of commercial scale accurately. The small scale studies or lab scale studies are shown in Fig. 12.10.

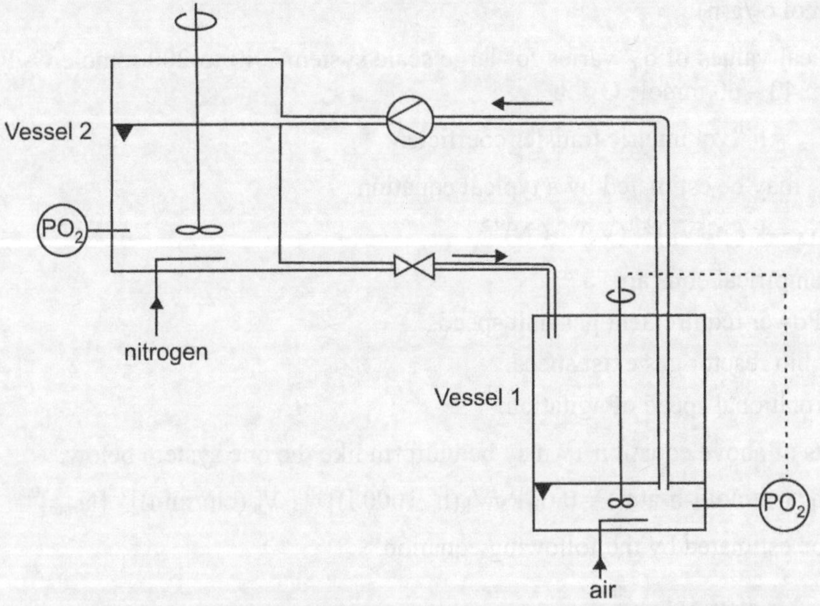

Fig. 12.10 Scaled down of experiments – an experimental set up.

And they represent the substrate and dissolved oxygen concentrations that are expected for a mixing time analysis or for process data of residence time distributions of commercial scale reactor. The basic difference between scale down unit and smaller reactor in the scale down unit is actual mimic of commercial equipment. While scaling down it may be noted that the heterogeneity in temperature may not be reproduced in smaller system. But the modifications in temperature control can be achieved by studying separately in each bioreactor. Such scale down reactors may be used to study the system response like growth rate, product formation and other changes in medium composition, introduction of modified production strains, O_2 and CO_2 tolerance and new antifoam agents etc. The other studies may include corrective protocols to be used in large scale system after carrying out simulation studies in small scale reactors. Particularly like oxygen probe failure or compressor failure during different stages of bioreaction.

So small scale reactors from scale down process will be useful for mathematical models, scale up rules and pilot plant operations.

Considerations in Aeration, Agitation

For industrial scale bioreactors, oxygen supply and heat removal are considered as key parameters. Oxygen requirements depend upon the microorganism chosen.

The equation for oxygen uptake rate (OUR) can be written as,

$$\text{OUR} = X\, q_{o2} = K_{La}(C^* - C_L) \qquad \ldots\ldots(16)$$

Where

Q_{o2} = specific uptake rate of oxygen

(mol o_2/g-h)

The typical values of q_{o2} varies for large scale systems, 40 to 200 mmole O_2/l-h), and for most system 40 – 60 mmole O_2/l.h

$K_{La} = h^{-1}$ volumetric transfer coefficient

Now K_{La} may be estimated by a typical equation

$$K_{La} = k\, (\, \rho_g/V_R)^{0.4}\, (V_s)^{0.5}\, (\,N)^{0.5} \qquad \ldots\ldots(17)$$

K = empirical constant

ρ_g = Power requirement gas unit speed.

V_R = bio reactor gas exist speed.

N = rotational speed of agitation.

The units of above equation 17 may be uniform like the one system below.

$$K_{La}\ (\text{m mol/l-h-atm}) = 0.6\ [P_g/V_R(h_p/1000\ 1)]^{0.4}\ [\ V_s\ (\text{cm/min})]^{0.5}[N_{rpm}]^{0.5} \quad \ldots\ldots(18)$$

P_g will be estimated by the following equation

$$P_g = K\left(\frac{P_u^2.N.D_i^3}{Q^{0.56}}\right)^{0.45} \qquad \ldots\ldots(19)$$

Where

K = constant based on reactor geometry.

P_u = Power required in the un gassed bioreactor.

D_i = impeller diameter

$$Q = \text{aeration rate} = \frac{\text{Volume of gas supplied / (min)}}{\text{Liquid volume of reactor}}$$

Some typical values of q_{o2} are given in Table (12.1)

Table 12.1 Typical respiration rates of microbes and cells in culture

Organism	q_{O_2} (mmol O_2/g dw-h)
Bacteria	
E. coli	10-12
Azotobacter sp.	30-90
Streptomyces sp.	2-4
Yeast	
Saccharomyces cerevisiae	8
Molds	
Penicillium sp.	3-4
Apergillus nigher	ca. 3
Plant cells	
Acer pseudoplatanus (sycamore)	0.2
Sacchrarum (sugar cane)	1-3
Animal cells	$0.4 \dfrac{\text{mmol } O_2/1-h}{10^6 \text{cells/ml}}$
HeLa	
Diploid embryo WI-38	$0.15 \dfrac{\text{mmol } O_2/1-h}{10^6 \text{cells/ml}}$

However the correlation may not be suiting to Newtonian systems or Non–Newtonian or highly viscous systems and effects of medium components on K_{La} in neglected. The bubble size and liquid film resistance around the gas bubble may be affected by the presence of salts and surfactants. Also the oxygen solubility (c^*) may also be affected. K_{La} and C^* may be affected by temperature and pressure also. However, C_L may be maintained above the initial oxygen concentration but a low enough value to provide good oxygen transfer. The measurable parameter K_{La} may be predicted by four approaches 1. unsteady state, 2. steady state, 3. dynamic and 4. sulfite test. Initially the new reactor can be filled with pure water or medium to measure C^*. N_2 is sparged to remove oxygen.

During unsteady state method air is introduced and change in dissolved oxygen (DO) is monitored till the solution gets saturated.

Here we have

$$\frac{dC_L}{dt} = K_{La}(C^* - C_L) \qquad \qquad .....(20)$$

$$-\frac{d(C^* - C_L)}{dt/(C^* - C_L)} = K_{La} \qquad \qquad .....(21)$$

$$\ln(C^* - C_L) = -K_{La}t \qquad \qquad .....(22)$$

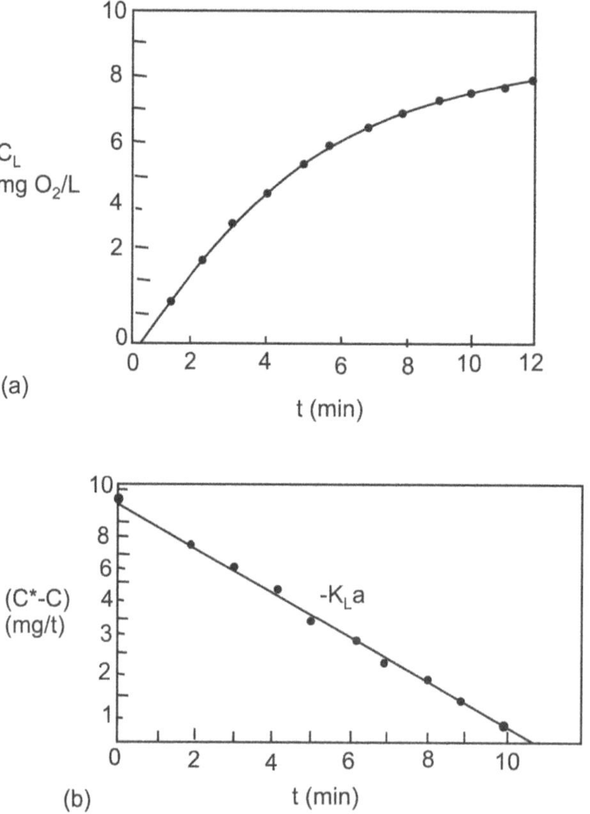

(a)

(b)

Fig. 12.11 (a) Typical data for unsteady state accumulation of oxygen in large stirred tank
(b) Assuming $C^* = 8.65$ mg O_2/ml a value of $k_{La} = 11.5$ h^{-1} is determined.

In Fig. 12.11 the plot of log $(C^* - C_L)$ versus time is used to estimated by K_{La}

Sulfite Method : The sulfur in sulfite (SO_3^-) is oxidized to (SO_4^{2-}) in zero order reaction in the presence of CO^{2+}. So C_L approaches zero as the reaction is rapid.

The sulfite formation rate is monitored and this is proportional to the rate of oxygen consumption (1/2 mol of O_2 is consumed to produce 1 mol of SO_4^{2-})

So

$$\tfrac{1}{2}\ dC_{so4}/dt = K_{La}\ C^* \qquad\qquad .....(23)$$

Where C_{so4} = concentration of SO_4^{2-}

C_{so4} and C^* should be expressed in terms of moles.

C^* = Oxygen solubility depends as medium compositions temperature, pressure and can be measured separately.

Now

$$K_{La} = \frac{1}{2}\frac{dC_{So_4}/dt}{C^*} \qquad \dots\dots(24)$$

K_{La} value may be over estimated by sulfite method.

K_{La} may be estimated by steady state method and the whole reactor may be used as a respirometer. Oxygen concentration in gas exit streams and C_L estimation may be accurately measured.

Mass balance on O_2 can be used to measure OUR

$$K_{La} = \frac{OUR}{C^* - C_L} \qquad \dots\dots(25)$$

Considering the actual bioreactor to be ideal, OUR can be estimated by off time measurement of a sample culture in respirometer. In large bioreactors, the gas in sparged under high pressure. And C^* is proportional to pO_2. This depends on total pressure and fraction of the oxygen. Near the sparger pO_2 will be higher than at the exit, due to higher pressure and decrease in fraction of oxygen due to consumption by respiration. Log mean value of C^* based on pO_2 at the entrance and exit would be taken, in bubble column. In a bioreactor perfectly mixed, exit gas steam composition should be the same as bubbles dispersed anywhere in the tank so the C^* would be based on pO_2 at the exit. However, in an actual tank, the knowledge of the residence time distribution of gas bubbles is necessary to estimate a volume averaged value of C^*.

12.5 Dynamic Method

This method like steady state method uses a bioreactor with active cells. It requires only a dissolved oxygen (DO) probe and a chart recorder rather than off gas analyser and DO probes as required in steady state method.

So the equation for DO levels is given as

$$dC_L/dt = OTR - OUR \qquad \dots\dots(26)$$

OR

$$dC_L/dt = K_{La}(C^* - C_L) - q_0.X \qquad \dots\dots(27)$$

In this method air supply is shut off for a short period, less than 5 min. and then turned back on. The result is shown in Figure.12.12. K_{La} will be zero when gas is off due to lack of gas bubbles.

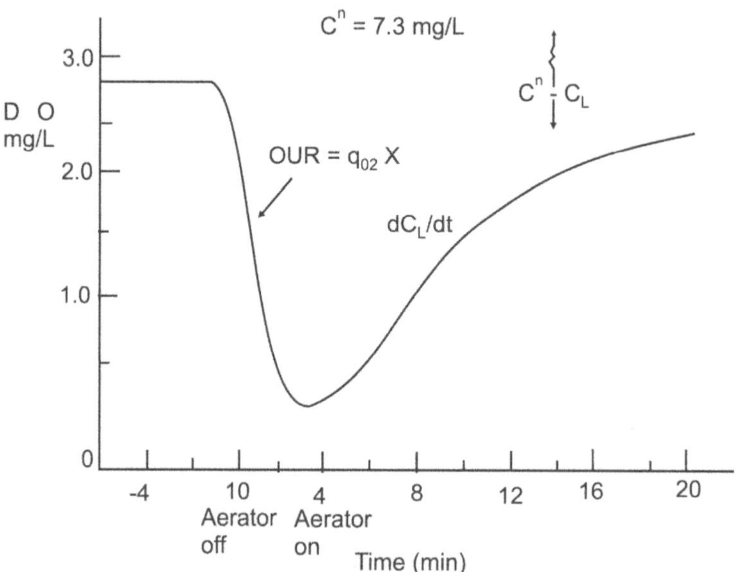

Fig. 12.12 Example of response of DO in fermentor when stopping and then re starting air flow.

So

$$dC_L/dt \;=\; -q_{o2}\,X \qquad\qquad .....(28)$$

The slope of the descending curve will give OUR or $q_{o2}\,X$. However the lowest value of C_L obtained in this experiment must be above the critical oxygen concentration and q_{o2} is independent on C_L. Also it is expected that complications may be due to measurement lag in DO probe. that is about 30 to 45 sec and the dissolution of oxygen from the head space gas into liquid which may be significant primarily in small reactors, as the ratio of surface areas for gas transfer is high compared to liquid volume . Now when the air sparging is resumed, K_{La} can be calculated by ascending curve. dC_L/dt values can be estimated from the slope of the ascending curve at various time points. The plot of (dC_L/dt + OUR) versus (C^* -C_L) gives K_{La} which is the slope of the line. The advantage of the dynamic method K_{La} is estimated under actual bioreactor conditions. The general principles of various methods for estimating K_{La} are same. OTR can be estimated from either correlations or experimental determined values of K_{La}

References

1. Abia S., A.E. Humphrey and N.F. Mills, (1973), "Biochemical engineering ' 2nd ed, Academic press, New York

2. Bartholomew, W.H, (1960) "Scale up of submerged fermentation" Adv. Appl. Microbiol. (Ed) Umbreit, W.W. Academic Press N.Y

3. Blanch, H.W and D.S. Clark "Biochemical Engineering" Marcel Dekker, Inc, New York, 1996.

4. Doran, P.M (1995) "Bioprocess Engineering Principles", Academic Press, San Diago,

5. Richards J.W (1961) "Studies in Aeration and Agitation" Progress in Industrial Microbiology

Review Questions

1. What is scale up ?

2. Write about the physical concept of scale up?

3. Describe biological concept.

4. Explain Power to unit value ratio.

5. Write about volumetric oxygen transfer coefficient.

6. Explain aeration and agitation.

7. Write sulfite method and dynamic.

CHAPTER 13

Bioreactors Instrumentation and Control

Introduction

The bioreactor control is an important part in maintaining optimum conditions in bio-product formation. A few parameter like, pH, dissolved oxygen and temperature are controlled in complex environment. The measurement of various ions will be Carried out by developing new probes and techniques for incorporating in different fermentation conditions. For example, oxygen and CO_2 are measured by using new types of probes for estimating conditions during growth phase of microorganisms. The bio-reactor control systems which are essential for any typical bioprocess which need to be developed, are discussed briefly here.

Measurements During Fermentation

The measurement of various parameters by using probes that extend into fermentation or bioreactor and lead to the contamination of such probes. It is therefore essential to have sterilizable probes that laid in side the bioreactors. They should withstand steam sterilization temperature, i.e., 121^0C similarly the probes also must be resistant to chemical in the case of the chemical sterilization which is not used frequently.

The various techniques available for monitoring physical environment is given in Table 13.1.

Table 13.1 Physical environment – monitoring and control .

	Parameter	Measuring devices(s)	Comments
1.	Temperature	Resistance thermometer or thermistor	Thermistors are used when small size and rapid response are required
2.	Pressure	Diaphragm gages	Pressure is regulated by a simple back-pressure regulator in gas exit line
3.	Agitator shaft power	Wattmeter or strain gages	Strain gages are used in bench or pilot-scale equipment
4.	Foam	Rubber –sheated electrode	Mechanical foam breakers are self-regulating; sensor is used to activate solenoid valve to release antifoam agent
5.	Flow rate (gas)	Rotameters or thermal mass flow meter	Position of rotameter float is converted to an electrical signal; controller manipulates flow valve

Table 13.1 Contd....

Parameter	Measuring devices(s)	Comments
6. Flow rate (liquid)	Magnetic-inductive flow meters or change in weight of additional vessels deter Mined by load cell	Magnetic-inductive meters are good for viscous fluids with high level of particulates; controller can manipulate flow valves.
7. Liquid level	Load cells to measure amount of liquid in vessel. Liquid height conductivity sensors; capacitane probes; ultrasound	Liquid height is a function of gas sparge rate and gas hold-up; foram can complicate measurement
8. Viscosity	Rotational viscometers	On-line measurement is difficult; broths with high solids content present special problems
9. Turbidity (to indicate cell	Photometer (either as a probe into the reactor or in a slip stream)	Many problems; fouling and interference from gas bubbles and suspended soli

These parameters are monitored at pilot and industrial scale bioreactors but turbidity and viscosity are not generally monitored at large scale level. The parameters are subjected to closed loop control system. The individual control loops can be integrated into an overall control package for large scale units but such systems are not usually in practice due to lack of accurate problems models. The key component concentrations are determined by various techniques as summerized in Table (13.2)

Table 13.2 Chemical environment – monitoring and control

Approach	Possible measuring Derived and compounds Monitored	Comments
1. Insertable probes	Redox electrodes; redox potential	Measurement fairly reliable; interpretation and use of information often difficult
	Ion-sensitive electrodes; NH_3, NH_4, Br, cd, ca, cl, cu, CN, F, BF_4, I. Pb, NO_3, Clo_4, k, Ag, Na, S, SCN	Cross-sensitivity major problem; drift and calibration can be troublesome; response time can be slow; sterilizability is often poor
	O_2 probes, either galvanic or polarographic pO_2	Measure partial pressure of O_2, not O_2 concentration; slow response drift, fouling, and recalibration during extended fermentations can be a problem; O_2 probes widely used and extremely important
	CO_2 probes; activity of dissolved CO_2	CO_2 is known to have important physiological effects, but relating these effects to dissolved CO^2 is difficult
	Flurescence probes; NADH	Commercially available to monitor intracellular concentrations; fouling is potential problem

Approach	Possible measuring Derived and compounds Monitored	Comments
2. Exit gas analyzers	Paramagnetic analyzer (O_2); thermal conductivity or long-path infrared analyzers (CO_2)	These instruments are specialized for measurements of these specific gases; important in fermentation balances sample conditioning important
	Flame ionization detector; low levels of organically bound carbon, especially useful for volatile organics such as ethanol or methanol	Flame ionization measures total hydrocarbon; automatic process gas chromatography equipment available to separate into individual compounds
	Mass spectrometer, O_2, CO_2, volatile substances (can also be used on liquid streams)	Highly specific, rapid, and accurate expensive has not been used for
	Semiconductors gas sensors; flammable or reducing gases or organic vapors	complex molecules; depends on gas phase for volatile compounds has been used in bioreactors primarily for ethanol measurement
3. Measurements from liquid slip stream	HPLC(high-performance liquid chromatography); dissolved organics; particularly useful for proteins; auto anlyzer	Highly specific; compound need not be volatile; response time is long (min); solids must be removed by filtration; membrane foulding can be a problem; guard columns must remove compounds that may foul columns; expensive; main value is to differentiate am ong closely related proteins or variants or a particular protein
	Mass spectrometers dissolved compounds that can be volatilized	Same instrument can be used also for exit gas analysis; rapid, specific, but expensive

The probes for measuring parameters like, dissolved oxygen and substrate levels, are to be placed in more turbulent zones to get uniform values. But it is not possible due to mechanical design considerations some times they may be placed at different locations in the bioreactor. To get uniform concentrations, various probes may be inserted, however, due to contamination of probes this is not suggested. If these probes are placed in turbulent zones, contamination on the probes, due to reactor contents may be reduced.

The fermentations or bioreactions take longer time i.e., 2-20 days for completion of a reactions. The probes inserted into the bioreactor, must also gives stable responses for the longer time. The other interesting problems would be the adherence of organics, proteins or microorganism present in bioreactor liquid on to the probe surface. So maintenance of sterility in such situations may be difficult.

However, certain systems like back flushing may be useful in removing such adhered contaminants.

The other type of measuring instruments include exist gas measurement instruments with mass spectrophoto meter. The costs of such robust instruments are available at affordable costs resulting in wider applications in exit gas measurements. However, the limitations would be that they can monitor volatile components of exit stream.

The dissolved solutes mostly proteins are measured by using HPLC (High performance liquid chromatography]. This method is being adopted, by many people though it is not required for routin analysis. The liquid slip stream are used for sample analyses and often the sample preparation takes quite some time. This results in delay of the response to the input stream. The contamination is other problem of concern as far as slip steam is concerned. The sample has to be further treated by filtration to remove cells and particulates. Interestingly, the slip steam reduces the time delay and useful in rapid determination.

For small growth experiments or off line measurements for intercellular metabolism NMR (Nucleus Magnetic Resonance) is frequently used. Although such measurement by online measures are in practice, the practical difficulties may be solved. Analytical tools like FTIR (Fourier Transfer Infrared) spectra are used, for spectral analysis, combined with optical flux for determination of cell mass in the presence of suspended solids.

13.2 Overview of Control of Fermentation

The bio-reaction or fermentation can be regulated by controlling the environment prevailing in the bioreactor. In order to properly regulate a bioreactor, it is necessary to understand the metabolic pathways and also proper instrumentation to moniter and control process parameters.

Biochemical engines have to focus their attention for evaluating the metabolic pathways and bioreaction mechanisms and also to properly control parameters by adopting appropriate instrumentation.

The bioreaction can be controlled by following different steps as given below

Step 1:   Carry out bioreactions in controlled environmental conditions

$\downarrow$

Develop

Step 2 :   Correlation between cellular control mechanism and environmental conditions

$\downarrow$

Step 3:   Reproduce environmental conditions by computer motoring, analysis and feed back control of bioreactors.

The above three steps have been followed to develop a meaningful and accurate system to control any biochemical reaction. With the advent of new control system and reaction logic contol the fermentations are being carried out by computer interfaced bioreactors.

Sensors for Monitoring Reactor Environment

The basic bioreactor environment to be controlled is based on two types of environment sensors, 1. Sensors for controlling physical environment 2. Sensors for controlling chemical environment. The physical and environment factors are represented in Table (13.3) and (13.4) respectively.

Table 13.3 Physical environment Factors

Systems involved in Controlling the Physical Environment
Temperature
Pressure
Foam
Flow rate (gas and liquid)
Turbility
Viscosity

Table 13.4 Chemical environment factors.

Existing systems
PH
Redox
Dissolved oxygen
Dissolved car dioxide
Oxygen in exit gas
Carboin dioxide in exit gas
Precursor level and feed rate
Sugar (carbon) level and feed rate
Protein (nitrogen) level and feed rate
Potential systems
Mineral ion level
Mg^{2+}, K^+, Ca^{2+}, Na^+, Fe^{2+}, SO_2^4, PO^3_4
RNA
DNA
NAD, NADH
ATP, ADP, AMP

Most of the physical environment factors can be monitored by available sensors. As far as chemical environment factors are concerned, the adequate montoring systems are available for pH, redox, dissolved oxygen and exist gas only. The sensors for other chemical environment factors are being developed by companies like Forbes Marshall and Orbit International etc.

13.3 Sensors for Physical Environment

Temperature:

This is monitored by 1. Thermometer bulbs (Hg in steel) 2. Metal resistance thermometers 3. Thermo couples 4. Thermistors

The thermistor coupled to a cascade control system may also be used to regulate temperature.

Pressure

Diaphragm gage may be used to detect pressure. Back pressure regulator valve in the gas exit line may be used to control pressure in bioreactors.

Shaft Power of Agitators

Usually the watt meter attached to agulators is used to measure power of agitator shaft is commercial bioreactors.

However bench scale bioreactors, the power measurement is carried out by 1. Torsion dynamometer 2. Strain gage.

The bearing and seal friction losses are included in dynamometer system for measuring agitator power in bench scale bioreactors. So the strain gage system is used due to its accurate measurements. The balancing strain gages are mounted on the agitator shaft inside the bio-reactor. The gages lead wires are taken out of the reactor through an axial hole in the shaft. Electrical slip ring arrangement takes the electrical signal from their rotating shaft. The details of this arrangement are shown in Fig. 13.1 and 13.2.

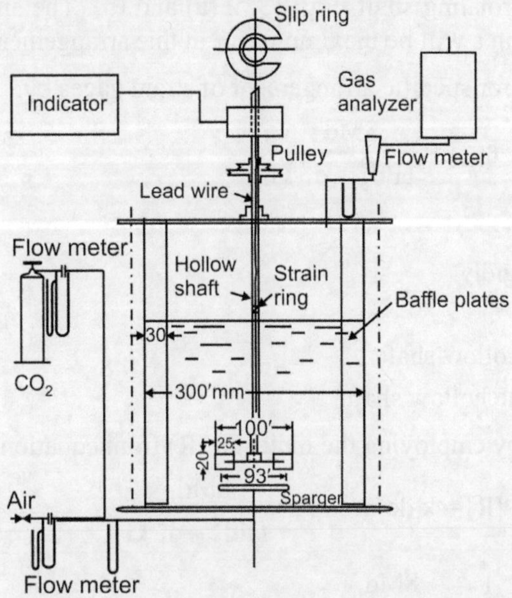

Fig. 13.1 Fermentor with strain gauge system.

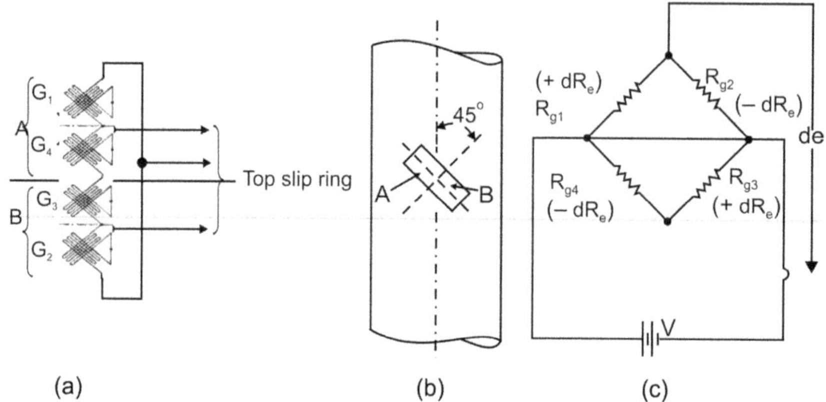

Fig. 13.2 Strain gauge monitoring arrangements.

Four strain gauges are connected as shown in Fig. 13.2 (C) and gauge is exposed to longitudinal stress σ and results in a strain ± ε. The value of potential (de) generated between the terminals of the bridge in each gauge is given by,

$$de = VK_{st} \qquad \qquad .....(1)$$

Where

V = electrical potential applied

K_s = gauge factor.

The gauges are mounted on the shaft and they are fitted to 45^0 C to the axis in order to measure the torque of a rotating shaft. Fig. 13. 2 (a) and (b). The angle of 45^0 C is selected because the shearing strain ε will be maximum ε_{max} at this arrangement .

ε_{max} can be estimated for specific arrangement of strain gages by.

$$\varepsilon_{max} = \frac{8Md_o}{\Pi\left(d_o^4 - d_i^4\right)G} \qquad \qquad .....(2)$$

Where

G = Modules of rigidly

M = torque

D_0 = outer dia of hollow shaft

D_i = inner dia of the hollow shaft.

de'; can be obtained by employing the indicator (R) from equation (1) and (2)

$$[R] = k\, de = kVk_s \frac{8Md_0}{\Pi\left(d_0^4 - d_i^4\right)G}$$

$$\frac{1}{\alpha} \frac{8Md_0}{\Pi\left(d_0^4 - d_i^4\right)G} \qquad \qquad .....(3)$$

Where

K = conversion factor which varies depending on the specific indicator used.

So as to sterlize the gages at 120^0 C the gages may be coated with polyster material. The agitator shaft power has to be recorded continuously during fermentation and this is obtained by multiplying observed value of M by angular velocity of the impeller.

Foam

Foam is the another parameter that need to be controlled by various means like

1. Mechanical foam destruction 2. Anti-foam agents chemicals
2. Both by mechanical and chemical means.

13.4 Mechanical Foam Breaking

These types of systems are self regulating and they do not require any sensors. An example for such is centirifugal defoamer. However, for severe foams, chemical antifoams agents are also to be added to reduce the foam.

Mechanical and Chemical Agents

An anti-foam system is shown in Fig. 13.3.

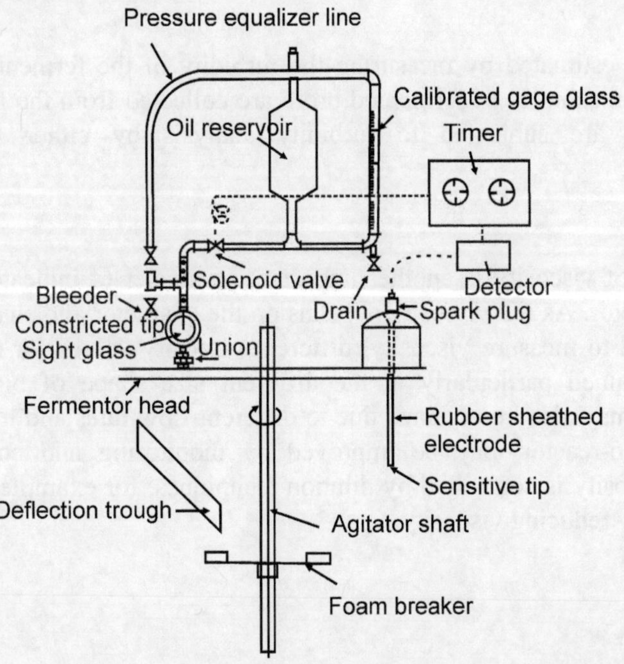

Fig. 13.3 Ancillary equipment to control foaming in fermentation both.

When the foam comes in contact with a rubber sheathed electrode, an electrical circuit actuals a solenoid valve to allow the passage of a sterile anti-foaming agent in to the bio-reactor. Antifoaming agent is dispersed uniformely over the foaming surface with the help of a deflection trough as shown in Fig. 13.3. In addition to this centrifugal effect of foam breaker also helps in reducing the foam density. The time fitted in the circuit to the solenoid value will control the addition fo antifoam agents into the bioreactor.

13.5 Chemical Anti Foam Agents

Silicone emulsions anti-foams can be easily sterilized and they control foam to great extent. Since silicone emulsion anti-foam agents are costly, oil based anti-foams are utilized in industrial fermentations. These anti-foam agents should be exposed to temperatures upto 160 °C for 2-3 hr for maintering sterile conditions

13.6 Gas and Liquid Flow Reaction

Gas flow rate can be measured by a variety instruments. Rotameter, the flow meters are normally employed to measure the flow rates. The position of the float of the rotameter will be converted to electrical signal through a capacitance or resistance principle. The control takes this signal to regulate the valve of the gas line.

Electromagnetic Flowmeters are used for the Measurement of Liquidflow Rates.

13.7 Turbidity

The cell growth is estimated by measuring the turbidity of the fermented broth inside the bio-reactor. Random samples of fermented broth are collected from the fermented broth and then the samples are subjected to turbidity analysis by either turbidity meter or Spectrophotometer.

Viscosity

The measurement of viscosity is another important parameter to indicate the microbial cell growth in a bioreactor. As the viscosity depends on the character and shape of the flow field in the devices used to measure viscosity, different viscosity values for the same fermented broth may be obtained particularly in the different size shape of bioreactors. Different viscosity readings may also be obtained due to different flow rates and mycelial shapes. The performance of bio-reactor may be improved by monitoring and controlling viscosity. Generally, the viscosity is controlled by dilution techniques, for example, by adding 15% of nutrient and thereby reducing viscosity.

Chemical Sensors

pH

Combined glass reference electrode may be used for the measurement of pH. Sterlizable pH electrode is shown in the Fig. 13.4.

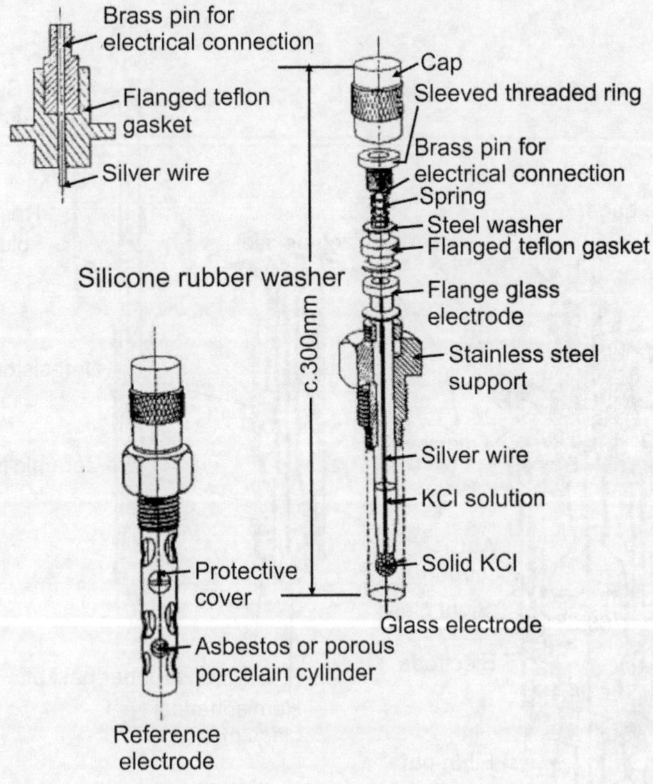

Fig. 13.4 Sterilizable pH electrode .

The half cell of glass electrode was composed of Ag/Agcl saturated with solid Kcl . The glass electrode and the half cell reference electrode contains the same material, where an asbestos or porcelin cylinder is used as junction material. The glass and reference electrode are mounted in Teflon gaskets and silicone rubber washers to ensure good insulation. The electrodes internal resistance is $300 - 500$ MΩ. However, both electrodes are protected by steel sleeves having several holes to allow free passage of the fermented broth. These electrodes could be used for 200 hrs without replacing KcL and during this period, electrodes are sterilized for three times with steam.

Another type of pH electrode is shown in Fig. 13.5.

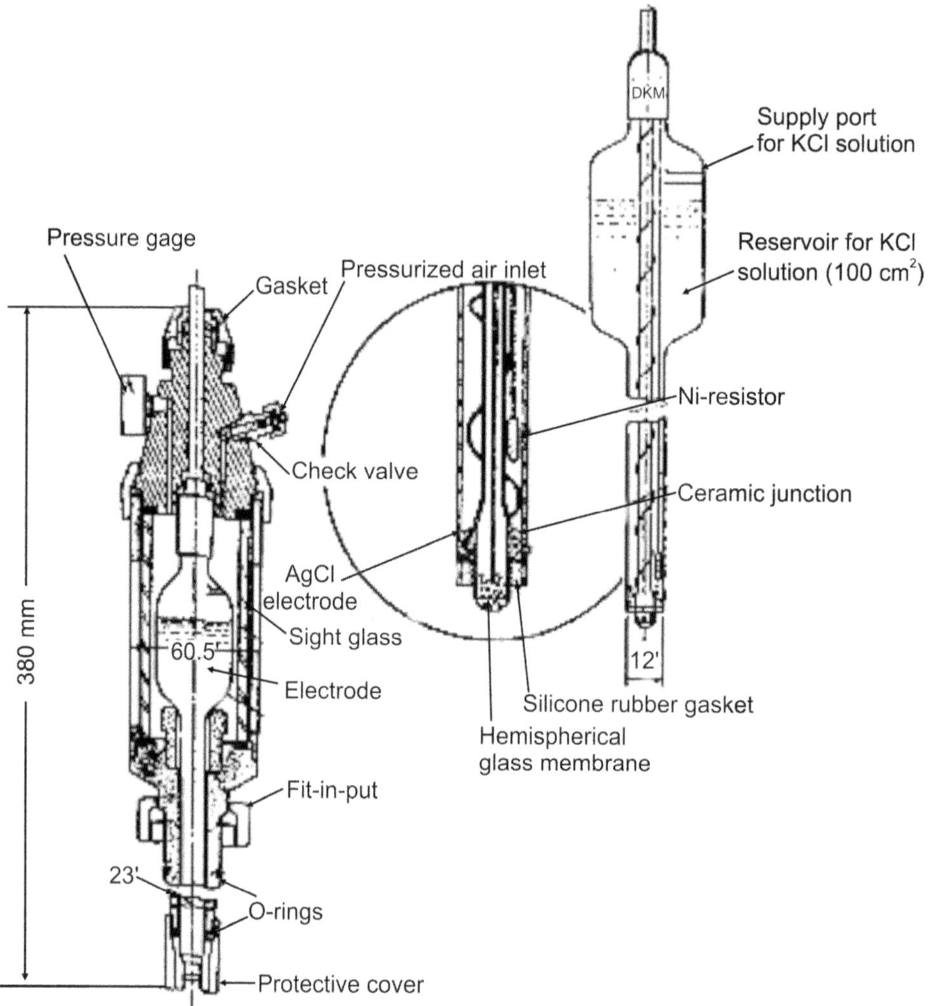

Fig. 13.5 "3 – in – 1" pH electrode.

This type of 3 in 1 electrode is used in fermentation industry. Thermal compensator, reference electrode and glass electrode are assembled to form single pH electrode. This glass electrode is characterized by small flow of KCL solution in the reference electrode into the broth through the ceramic junction at a flow rate of about 0.2 ml/ hr at 1.0 kg/cm^2 pressure difference between the inside and outside the electrode. The junction surface to broth of electrode is always kept clean and a stable potential is maintained between electrode and broth, due to the forced flow of KCL. This electrode maintains its performance in terms of zero point shift (with $\pm$ 15 mV) rapid response to pH (with in 1- 2 min) and mV/pH variations 98 – 99 % even after 50 sterilizations at 120^{0}C for 30 minutes.

Redox

In aerobic fermentations response to redox exist, as the oxygen is a growth regulating substrate. Also the dissolved oxygen interacts with the redox system. Mostly the redox measurements are made with platinum – references electrode systems.

13.8 Dissolved Oxygen

In place of steam sterilizable dissolved oxygen sensors are available in the market, to use in the bioreactor systems. A typical oxygen sensor Au (cathode) – Ag (anode) is shown in the Fig. 13.6.

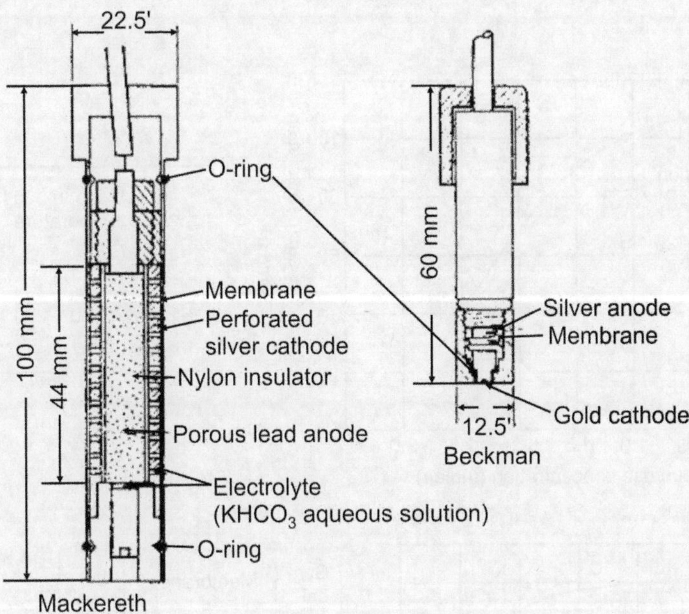

Fig. 13.6 Beckman and Mackereth.

A membrane covered electrode is immersed in the liquid medium and oxygen is supplied from outside the electrode as the oxygen is reduced at the cathode surface. So the oxygen is reduced at the cathode as given below.

$$O_2 + 2H_2O + 4e \rightarrow 4OH^-$$ …..(4)

Partial pressure profile of oxygen in the oxygen sensor in the steady state is shown in Fig. 13.7.

I. Resistances to oxygen flux are indicated as R_L, R_D, R_m, R_o^1 and R_E. In case the membrane is tightly attached to the cathode surface, then the values of R_o^1 and R_E would be assumed to be negligible. Due to rapid electrochemical reaction, the value of P_a at the cathode surface is equated to zero and R_D, the resistance at the medium-membrane interface would also be assumed to be zero.

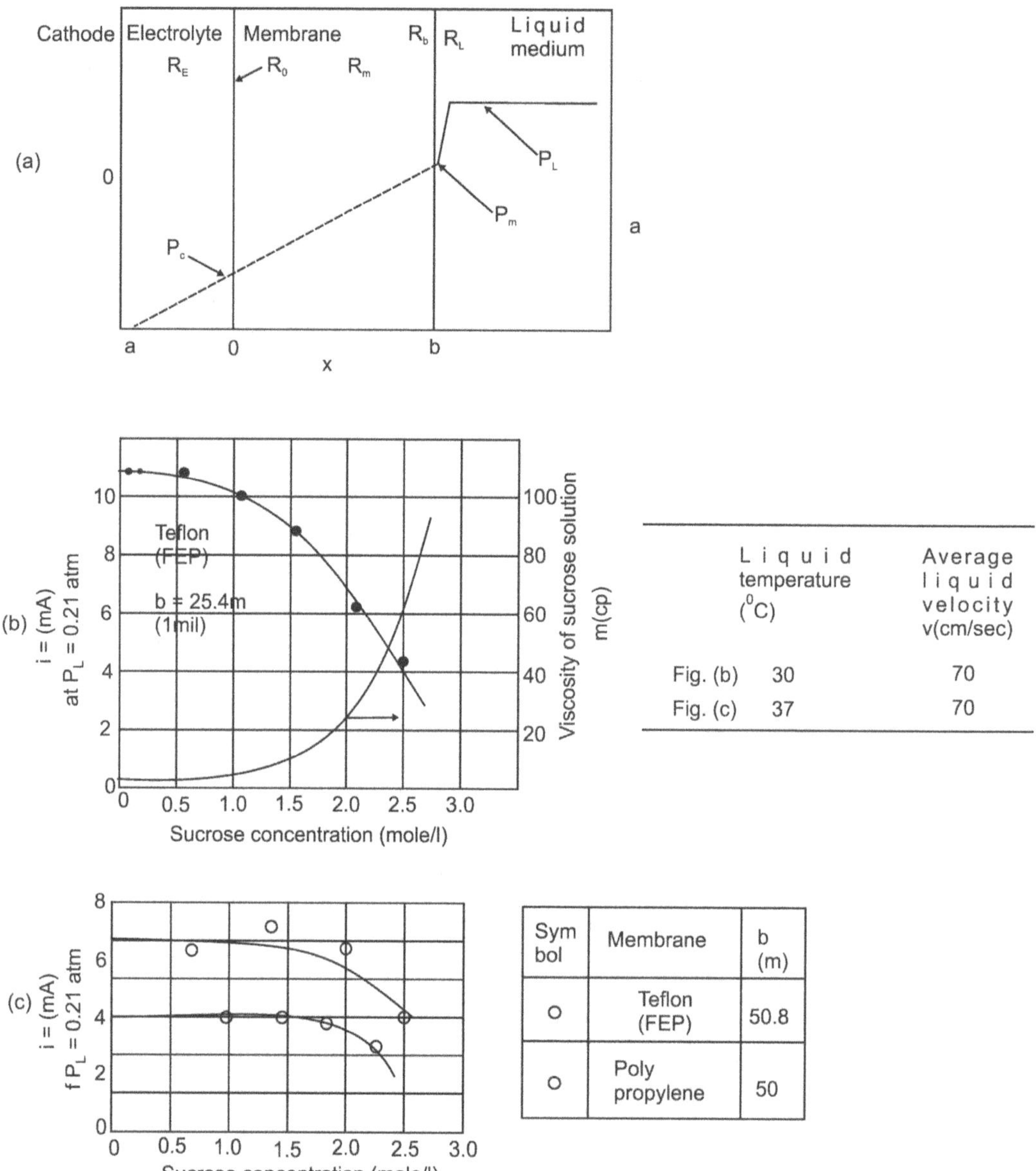

	Liquid temperature (^{0}C)	Average liquid velocity v(cm/sec)
Fig. (b)	30	70
Fig. (c)	37	70

Symbol	Membrane	b (m)
○	Teflon (FEP)	50.8
○	Poly propylene	50

Fig. 13.7 Principle of membrane covered oxygen sensor (a) Oxygen partial pressure profile in steady state (b) Example of steady sate current i_∞, decreasing with an increase of liquid viscosity. (c) example to correct previous decrease of i_∞ by replacing membrane.

In the steady state the oxygen flux n_{o2} is given as

$$n_{o2} = K_L (P_L - P_m) \qquad\qquad .....(5)$$

$$= K_m(P_m - P_o) \qquad\qquad .....(6)$$

$$= (P_m/b) (P_m - P_0) \qquad\qquad .....(7)$$

$$= K_0(P_L - P_o) \qquad\qquad .....(8)$$

Where n_{O2} = flux of oxygen

$K_L \left(\dfrac{1}{R_L} \right)$ = oxygen transfer coefficient in liquid film

$K_M \left(\dfrac{1}{R_m} \right)$ = oxygen transfer coefficient of membrane.

$K_o \left(\dfrac{1}{R_o} \right)$ = overall oxygen transfer coefficient

P_M = oxygen permeability of membrane

b = membrane thickness

P = oxygen partial pressure.

i_α, steady state current of the sensor in circuit is given as

$$i_\alpha = n \, F \, A \, n_{o2} \qquad\qquad .....(9)$$

i_α = steady state current

n = number of electron per molar unit of the reactions

F = Faraday's constant

A = Surface area of cathode

So as to operate the sensor with a view to keep i_α insensitive to physical conditions of liquid medium (viscosity) it is understood from equation (5) to (9) that the resistance.

$R_L = \left(\dfrac{1}{K_L} \right)$ of this liquid film outside the membrane, should be minimized as a fraction of

the overall resistance R_0, to the oxygen flux ($R_L/R_0 \leq 0.05$) with the increase of liquid viscosity, i_α value deteriorates as shown in Fig. 13.7(b).

The sensor will become less sensitive to the liquid viscosity upon replacement of Teflon membrane with a thicker one, which has a lower oxygen permeability. A sensor corrected for liquid viscosity becomes less sensitive to change in oxygen concentration in medium time and longer response time prevails.

The response of membrane covered sensors is normally low, for instance the 90% response is of the order of 10 – 100 sec. These sensors only give as qualitative picture of the short term oxygen fluctuations though average oxygen levels are measured adequately. The membrane covered sensors must be temperature compensated with a thermistor circuit that electrically matches of the temperature behavior, compensated as these are sensitive to temperature fluctuations.

Dissolved oxygen may be measured by tubing method. This contains a passing stream of inert gas like nitrogen or helium, through the coil of permeable tubing like Teflon or polypropylene and their measuring the amount of gas which is picked up by diffusion through tubing. This is not good monitoring system it has time delays 2 – 10 min Fig. 13.8.

Exist gas composition

Infrared analysers having long path or thermal conductivity devices, are used for exit gas composition monitering like that for CO_2. Oxygen is analyzed by paramagnetic analysers. The other simpler method would be to bubble exit gas through replenished water and then to monitor dissolved oxygen in replenished water.

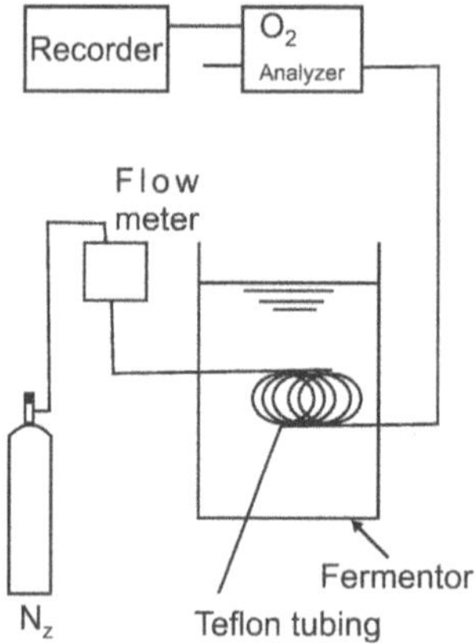

Fig. 13.8 A probe of Teflon tubing coupled to a gas analyzer for measurement of dissolved oxygen.

Computer controlled – Bioreactors

The bioreactors are complex in nature and control of such systems require data collection, analysis and feedback control. Laboratory scale and industrial scale bioreactor can be controlled by interfacing bioreactors /fermentators with advanced computers Dynamic optimization techniques have been evolved from the computer controlled bioreactor systems.

The computer controlled bioreactor is shown in the Fig. 13.9.

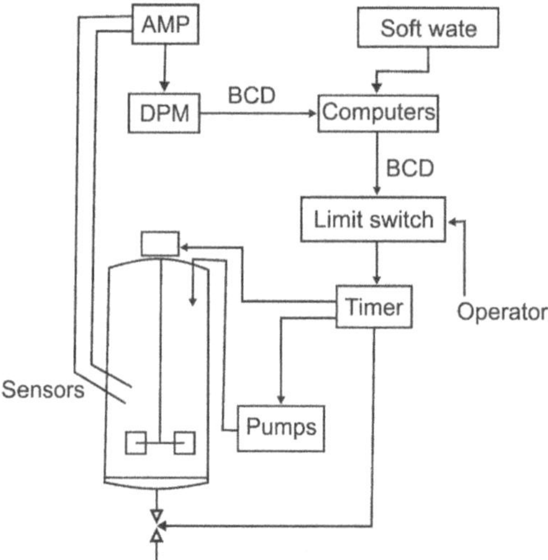

Fig. 13.9 General layout facilities for bio – reactor controlled.

Fig. 13.10 A typical computer controlled fermentor.

Bioreactor signals are amplified (AMP) and are displayed on digital panel meters (DPM). DPM converts the analog values to digital forms (BCD) and transmits them to the computer. The bioreactor and computer are linked via limit switches or set points on the individual controllers. The controllers then actuate pumps or values on a real time basis to alter the controlled variables of the system. The operator can determine the control set points as desired in order to assure flexibility.

Computer

The computer configurations in relation to its major function is shown in Fig. 13.11.

The online operation is performed by computer in three basic fields of functions.

1. Logging of data obtained through data aquisition system.
2. Data analysis on the basis of the algorithms
3. Control functions of valves and pumps to control certain process variables.

The choice of computer to perform data logging, analysis and control functions are based on

1. A suitable word capacity that assures the handling requirements of the whole system
2. An expandable memory that accommodates future needs, such as decision control loops
3. A large selection of peripheral equipment interface boards and soft ware library
4. Adequate field service ability.

A computer program as shown in Fig. 13.12 may be considered for developing a control system.

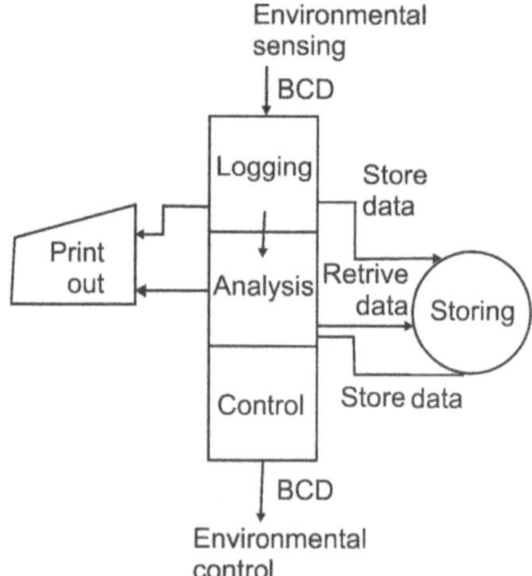

Fig. 13.11 Computer and its major functions.

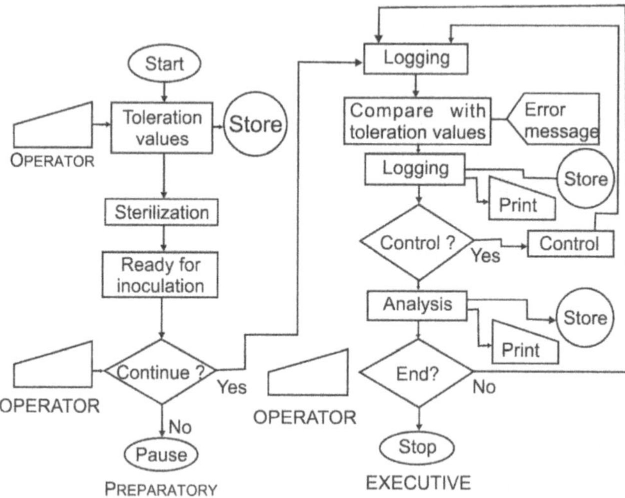

Fig. 13.12 Essential of computer program for fermentation process.

The program may include the following.

1. Preparatory phase for bioreaction/ fermentation.

2. Executive phase for bioprocess. During preparatory phase the following are considered.

   (i) Calculation of toleration values of bioprocess variables, that are introduced through operations console.

   (ii) Sterilization

The executive phase may include data lagging reduction and process control.

Data analysis

The computer for a typical bioprocess should provide an insight into the various aspects like, physiological and bio chemical conditions. The following parameters for data analysis are given in Table.13.5.

Table 13.5 Grouping of Algorithms for data analyses

Physical	Physico-chemical	Physiological	Biochemical
Agitator shaft power	Apparent viscocity	Oxygen uptake	Carbon balance
Rate of addition of ingredients	Power number	Carbon dioxide evolution RQ	Oganic energy yield
	Reynolds number		Cell mass based on Acid/base titration, Carbon dioxide evolution,
Broth volume	Flow characteristics Volumetric oxygen transfer coefficient		Carbon balance

The physico chemical characteristics may be combined to obtain information for scale up. A typical system shown in Fig. 13.13.

The program for detailed, real time study of carbon dioxide evolution and oxygen uptake of microbial cultures could be constructed due to applications of reliable gas analysis. The basic idea of the study of some physiological characterstics of cells with the help of computer is shown in Fig. 13.13. Carbon dioxide evolution and oxygen uptake by the culture can be used to obtain information on the actual respiratory activity of the culture. The data obtained regarding respiratory quotient and data on the carbohydrate metabolism as well as mass transfer coefficient can be correlated to control the process effectively.

The metabolic pathways can be well understood by studying respiratory activity and carbohydrate metabolism with the help of biochemical characteristics .The biochemical character sticks can be represented as shown in Fig. 13.14.

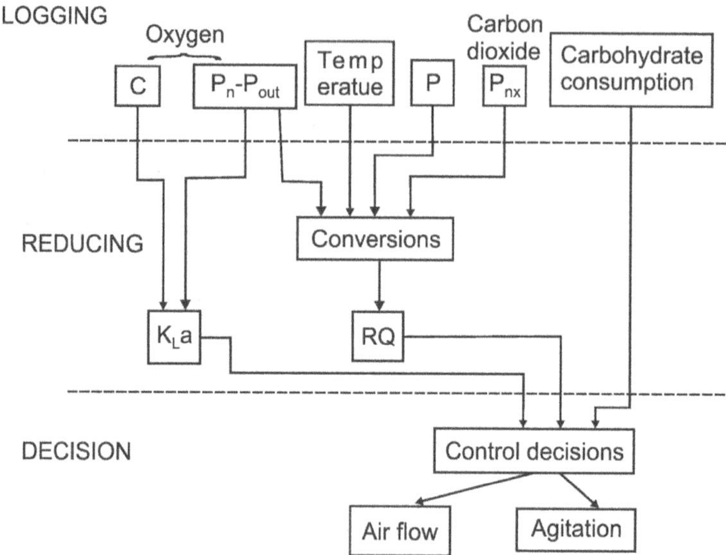

Fig. 13.13 Physiological characteristics.

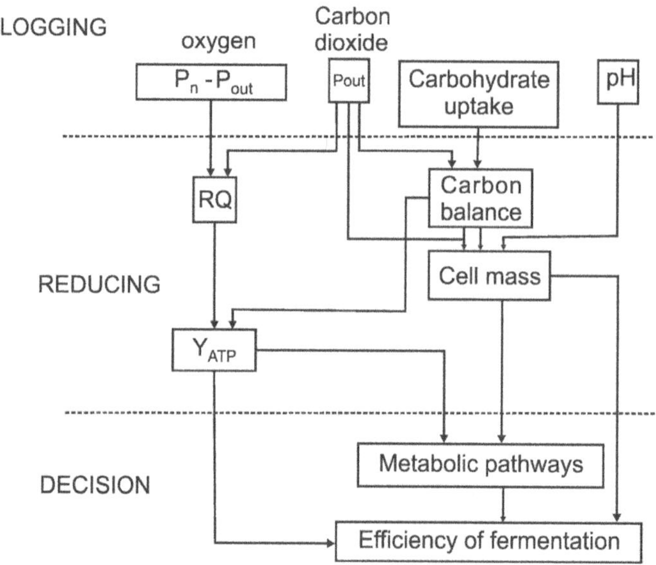

Fig. 13.14 Biochemical characteristics.

Y_{ATP} could be calculated by knowing the pattern of Carbohydrate metabolism and this can be related to formation of particular product or cell mass production.

References

1. Aiba, s. and Huang, S.Y, (1969) "Some consideration on the membrane covered electrode" J.Ferm. Tech (Japan).

2. D.G. Rao (2005), Introduction to Biochemical Engineering TMH, New Delhi

3. Hall E.A.H (1990) Biosensors, open University Press Buckingham.

4. Nelson, H. A., Maxon, W. D., and Elferdink, T.H. (1956), "Equipment for detail fermentation studies". Ind. Eng. Chem. (Vol 48).

5. Sobmouss, G.L. (1969) "Materials and methods in fermentation" Academic press New York.

Review Questions

1. Write about measurement during fermentation.

2. Describe overview of control of fermentation.

3. Explain sensors for Physical environment.

4. Write about chemical suspens.

5. Explain computer controlled bioreactors.

CHAPTER 14

Principles of Effluent treatment

Introduction

The effluent generated from various types of process industries, contaminate, fresh water ponds, revivers, lakes and even domestic waste water drains. This leads to human health problems particularly due to water born diseases. It is therefore, essential to treat the effluents before they are discharged into water bodies. Most of the industrial effluents can be treated by physical, chemical or biological treatment methods. It is found that biological treatment systems are cheaper than the available conventional effluent treatment methods. Biological treatments systems like aerobic treatment and anaerobic treatment will be discussed here.

The typical organic chemical waste water character sticks are given in Table (14.1)

Typical waste water characteristics of an organic chemical industries.

Table 14.1 Typical waste water characteristics of an organic chemical industry.

S.No	Parameter	Range (mg/l)
1.	pH	5.0 – 9.0
2.	BOD	300 – 30,000
3.	COD	1000 – 10,000
4.	TDS	500 – 15,000
5.	Suspended solids	50 - 1000
6.	Oil & grease	20 – 200

The industrial waste water emanated from, dairy industries, food industries, sugar industries, distilleries and most of the organic chemicals industries can be subjected to biological waste water treatment systems. The total treatment system can be divided into primary treatment system and secondary treatment system.

Primary treatment system contains (a) Physical treatment system (b) Chemical treatment system

14.1 Physical Treatment System

This includes, removal of grit material by bar screens, sedimentation, tricking filters.

14.2 Chemical Treatment System

This system contains neutralization of effluents by addition of either, the acid or alkali precipitation by the addition of alum, poly aluminium chloride etc. And treatment with chlorine for disinfection and some times treatment with hydrogen peroxide for oxidation of organic materials etc. The residual chlorine should be below 10 rpm.

14.3 Secondary Treatment System

This includes biological treatment systems like (a) aerobic biological treatment and (b) anaerobic biological treatment

14.4 Aerobic Biological Treatment

In this system the organic matter is degraded by the microorganisms, i.e., mixed culture in the presence of dissolved oxygen. Activated sludge process is considered to be one of the effective aerobic biological treatment system. B.O.D (Biological Oxygen Demand) of the effluent is reduced by about 80 – 90%. The flow diagram of a typical activated sludge process is given in the Fig. 14.1.

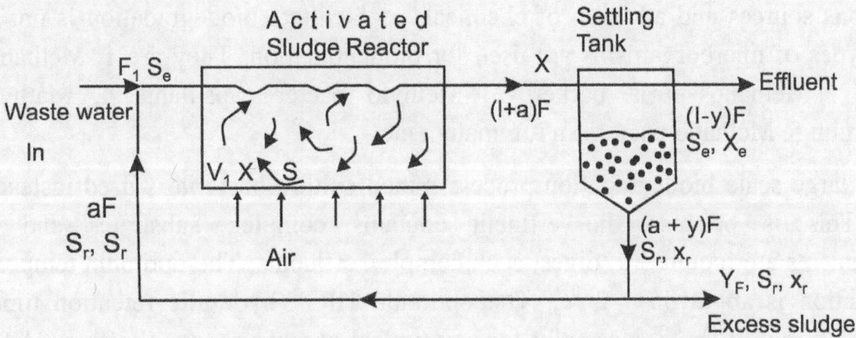

Fig. 14.1 A flow diagram of activated sludge process.

The air is passed in the aerobic bioreactor either by mechanical surface aerator or by diffused air i.e., by passing air through spargers by using compressed air. The sludge recycled to will be useful maintain MLSS (Mixed liquor suspended solids)in most of the effluent treatment. The dissolved oxygen level inside the aerobic bioreactor has to be maintained about 2-3 mg/L. The sludge from primary classifier is sent to sludge drying beds, whereas the sludge from secondary classifier can be used as manure since it is a biological sludge. The only limitation of the activated sludge process is the power consumed for continuous aeration of the aerobic reactors. It may be essential even to use generators to run aeration continuously without stopping. The aerobic bioreactor gives foul smell if the aerator is stopped for a long time.

14.5 Anaerobic Process

The biodegradation carried out in the absence of oxygen is known as anaerobic digestion process. Microorganism i.e., the absence of oxygen degrade the organic matter into methane and carbon dioxide. The microorganisms which convert organic materials into methane are known as methanogens. Most of the organic chemical industry effluents can be biodegraded to generate methane gas. Particularly the effluent generated from distilleries and breweries give foul smell if they are left to environment without any treatment. This problem can be solved by anaerofically digesting the effluents.

Bio gas Production

Dairy effluent, food industries effluents, distillery effluents and organic chemical industry effluents are potential sources for biogas generation. Experiments were also carried out to separate the biodegradation into 1. acidogenesis and 2. Methanogenesis. During acidogenesis, acetic acid is produced while during methanogenesis, methane is produced. However, it is difficult to separate into these phases. This type of biodegradation with separation of phases is known as biphasic methanation. Most of the effluents contain carbon, nitrogen and phosphorous sources and addition of chemicals to facilitate biodegradation is not required. Various types of microorganisms are used for biodegradation. They are 1. Methano coccus vanniele 2. Methanosapuria barkeri 3. Methylo bacter capsulants 4. Mathyl osinus trichosporuim 5. Methano bacterium ruminatum etc.

But in large scale biodegradation process mixed culture bacteria is used instead of pure culture. This is because the effluent contains complex substrates and different microorganisms are required to digest such complex effluents. The optimum temperature for biodegradation is about $35 \pm 2\ ^0C$ .The optimum HRT (hydraulic retention time) varies depending on the effluents. Normally for any typical organic chemical effluent HRT is 5-7 days but for Lingo cellulosic effluents the HRTs are longer i.e., $25 - 30$ days. The general reaction may be written as.

$$\text{Organic matter} \xrightarrow{\text{microorganism}} CH_4 + CO_2$$

The biogas generated from the food industry effluents would be in the order of 0.5-0.6 m^3 of organic matter digested. The bioconversion of the organic matter into methane is shown in Fig. 14.2.

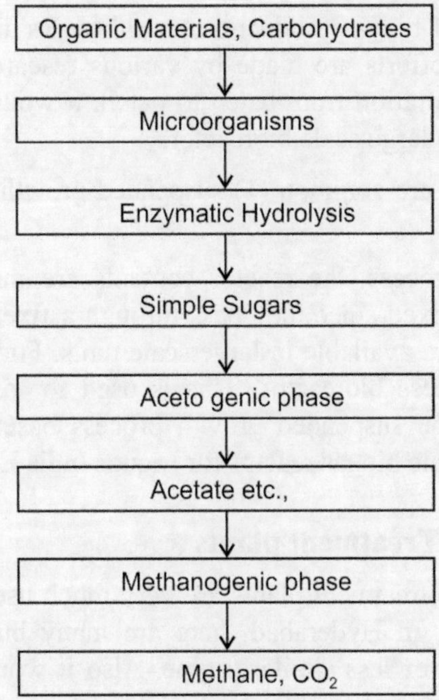

Fig. 14.2 Bioconversion of simple carbohydrates to methane CO$_2$.

The commercial gas is shown in Fig. 14.3

Fig 14.3 Commercial biogas Plant.

The approximate cost of biogas production would be Rs 1.5 – Rs 2.0 per m^3 of biogas production. Even through efforts are made by various researches to reduce HRTS due to variation of effluent concentration from batch to batch, it would be difficult to reduce HRT (hydraulic retention time) in large scale bioreactor.

Two types of bioreactor are common (1) suspended growth type bioreactors (2) attached growth type bioreactors.

In suspended growth process, the reactor contents are mixed and in attached growth process, contents are not mixed since they pass through a fixed bed of the microorganisms. Both types of bioreactors are available in large scale units. For example the Bacardi process is an attached growth process bioreactor. This is used for distillery effluent treatment at Andhra Sugars Tanuku. The suspended growth process based large scale bioreactors are many, for example, large scale biogas reactor for Liquor India Ltd., Hyderabad.

14.6 Common Effluent Treatment plants

These types of the effluents treatment plants are very much used for treating similar type of the effluents. For example, in Hyderabad there are many bulk drug manufacturing units which leave effluents more or less similarly type. Also it would be difficult for individual units to set up effluents treatment facility. So based on this concepts two common effluents treatment plants wore established in Hyderabad.

1.   Jeedimelta Effluent Treatment Plant Limited (JETL)
2.   Patancheru Effluent Treatment Plant Limited (PETL)

Both are really serving the needs of many small bulk drug industries in and around Hyderabad. The BOD and COD removal efficiency in aerobic treatment plants in below 50% mainly due to presence of high TDS (total dissolved solids). The depth of a typical aerobic bioreactors would be 1-2 m. The common effluents treatment plants are also facing operational problems due to variations in effluents concentration. Most of the Pharmaceutical industries discharge solvents also into their effluents, that are difficult to be recovered and these solvents are detrimental to the growth of microorganism.

However it may be suggested to pool up some pharmaceutical units, loads in small groups 2-3 units and build effluent treatment plants to reduce the pollution.

References

1.   Hammer, M.J., and M.J.Hammer, Jr, (2001), Water and Waste water Technology, 4th Edn Prentice Hall Upper Saddle River, EJ.

2.   Schroeder, E.D., (1997), Water and Waste water Treatment, McGraw-Hill Book Co., New York.

3.   Tehobanoglous, G., (1991), Waste water Engineering Treatment, Disposal and Reuse" (Metcalf & Eddy, Inc)., 3 Edn , McGraw-Hill Book, New York.

Review Questions

1. Write about biodegrading effluents.
2. Explain primary treatment.
3. Write about physical, chemical and biochemical methods of effluents treatment.
4. Describe aerobic biological treatment.
5. Explain common effluents treatment.

Bioprocess Engineering Medical Applications

The important area of the bioprocess engineering includes its application to medical sciences. The developments in molecular and cellular biology research over a period of time, have widened the scope of bioprocess engineering. Tissue culturing is one such area which needs attention for further development and gene therapy is another area, which is important for solving many individual specific diseases.

The development of tissues from the cells of healthy donors as in vitro will be a challenging area. These healthy cells can be transplanted to receptors to improve the damaged tissues. The example for such an application is, implantation of skin tissue and chondrocytes, in a damaged Knee for the production of cartillage tissue. Many other artificial tissues are being developed for catering the needs of human beings. These tissues that are in development include liver, pancreas, kidney, blood vessel, bone narrow etc. The artificial tissues produced can be used to test new medicines for pharmalogical and toxilogical testing. The use of animals used so far for carrying out pharmalogical testing may be reduced to great extent with the advent of tissue engineering. The living tissue is replicated in tissue engineering. The interaction of one cell with another and control of cellular differentiation mechanism should be known well prior to culturing of the cells in vitro. Bioprocess engineers and biologists must work together to develop reproducible type of tissue cells.

The formation of extracelluar matrix, the interaction of cells with one another and interaction with an artificial surface are of significance in developing tissue engineered constructions. Fibronectin and collagen are the extracelluar matrix proteins which are used for cell adhesion. Synthetic peptide sequences is 3 – 6 amino acids could be added to a surface at the end of a synthetic polymer to modify the synthetic substrates. The three dimensional organization of cells can greatly be, altered due to the difference in the strength of cell substrate and cell-cell adhesion. Various types of tools are being developed for attaining the improved performance during the formation of tissue for constructions. Commercial production of transplantable organs like liver etc; will be the future trend in tissue culture technology.

Tissue Culture Process

The classic examples of tissue culture are skin and cartilage and they represent novel developments achieved so far in this field.

Tissue Culture Skin

The important application of tissue engineering would be the skin tissues for replacement needed for patients with burn injuries. The tissue culture is gaining importance due to certain problems prevailing in cadaver skin, and animal products. In the case of cadaver skin it is not desirable due to potential disease transmission, possible immunological rejection, inflammation of woundbed, high cost and also limited availability. Also the other problem would be animal products which are not desirable due to immunological reactions. The available choice would be the skin taken out of the freezer used by the physicians, for examples, TranscyterTM, a human fiber blast derived temporary skin substrate. Routine circumcision discards are the sources of human neonatal foreskin, fibroblasts for making an artificial human skin. The fiborlast cells obtained in this manner are very much useful since they are not rejected by the body but also they can be replicated. They can be expanded in numbers in a straight forward ways. The method would involve the enzymatic digestion of donor tissue and seeding on to a three dimensional bioresorbable polymer scaffold. The cell growth secretion of proteins and extracelluar matrix material could be promoted in a bioreactor by placing cell – seeded scaffold of course the physiologically similar environment would be provided in a bioreactor. The three dimensional tissue construct could be made due to the cells and extra cellular matrix material with metabolically active cells which is functionally similar to natural skin and helps in healing of wound. The essential parameters like pH, temperature, oxygen level, nutrient supply, waste removal and fluid dynamics are controlled in bioreactor while maintaining desired tissues. The products produced would have uniformity and would have reproducible quality. A bioreactor is required for product growth and also cells as final packing during manufacturing process. The schematic process of production of skin substitute is shown in Fig. 15.1.

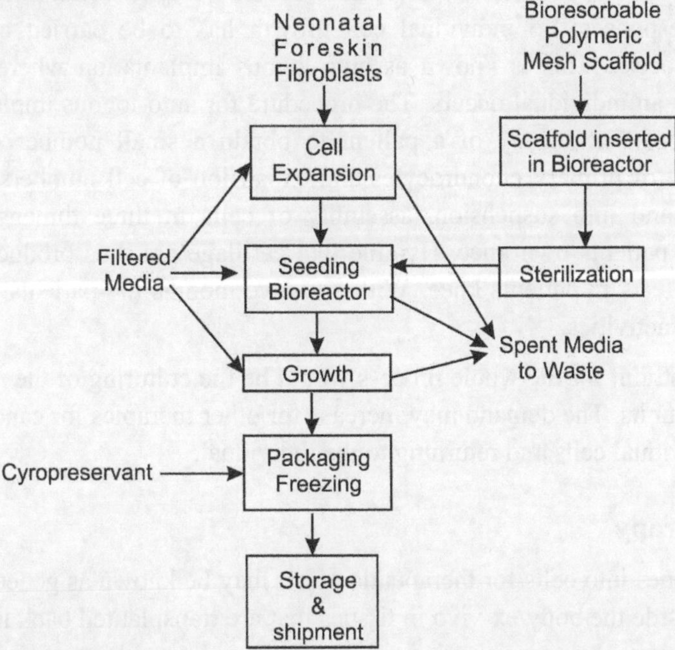

Fig. 15.1 Conceptual process diagram for production of a human skin substitute.

After assembly the system, is kept sealed for maintain sterility. The bioreactor is made up of maintaining sterlity, The bioreactor is made up of plastic material and large enough to hold a polymer mesh about – 13 by 19 cm. The bioreactor about a 12 in number are mainly folded together and are operated in parallel to obtain design size of product. After stylization it is opened for cell seeding and nutrient replenishment only. The cryopreservative is fed to the bioreactor module after three weeks of desired growth, will be packed frozen, stored and shipped after scaling the inlet and outlet tissues of bioreactor. The physician uses such products after thawing. FDA and GMP requirement will be followed during the manufacturing process.

15.2 Culturing of Chondrocyte

The common problems normally people face are the problems related to bones. The tissues get damaged particularly near the knee due to regular wear and tear. So their tissue layer found in the ends of the bones are articulate or hyaline cartilage. Chondricytes namely collagen fibers, proteoglycans and cells constitute cartilage. A very few fibrils cells are present 1% by volume but cartilage is formed due to synthesis and release of other compounds by these cells. The cells are no longer differentiated and do not make normal hyaline matrix proteins when chondrocytes are cultured as two dimensional or monolayer cultures when cultured in three dimensions such as suspension in agarose or in vivo, the cells redifferentiate and begin to manufacture hyaline like matrix. It is beneficial for patients to supply their own cells for expansion in number before being infected into the knee due to immunological responses. So individual cell growth has to be carried out in a separate manufacturing process. This is known as auto-logous implantation where the supply and reuse of cell from an individual occurs. The procedure for auto-logous implantation involves series of operations like biopsy of a palient to obtain a small number of chondrocytes, monolayer culture of primary chondrocytes and expansion of cell numbers, release of cells from monolayer and into suspension, assembly of cells in three dimensions and finally injection into the patients own knee. Hyaline like cartilage are thus produced by implanted cell and fill in defects in patients knee. Within twelve months the patients can resume their vigorous physical activities.

The basic constraint for the whole process would be the culturing of the tissues separately in manufacturing units. The demand may increase for other therapies for cancer treatment etc, for culturing individual cells and returning to the individual.

15.3 Gene Therapy

The transfer of genes into cells for therapeutic effect may be known as genetherapy. This can be carried out outside the body ex-vivo in tissues that are transplanted back into the patient or in a delivery vehicle such as virus in vivo where genes are injected into the patient.

Gene therapy is closely related to the metabolic engineering. The metabolic engineering for humans is more complex than the metabolic engineering of a single cell. This gene therapy has to be carried out carefully and should be added in right proportions in right time. The gene therapy is not much successful as the approaches are qualitative and trial and error in nature. The gene delivery may be carried out in many ways and involve viral vectors, use of naked DNA and liposome or particle mediated gene delivery.

Bioprocess technology would be required for the production of viral vectors.

15.4 Retrovirus

The virus vectors for gene therapy are retroviruses which are derived from a wild type of virus that infects mice. Retrovirus are encapsulated in a lipid bilayer membrane and may be known as enveloped viruses.

Retrovirus can be produced with much difficulty and demands good skill. These are produced by a two part system comprising a packaging cell line and the recombinant vector. Since recombinant virus cannot replicate itself, the two part system is required. A wild type virus is taken and essential verial genes are deleted and they are replaced with therapeutic genes of interest. The recombinant DNA is derived from such wild type virus. Reverse transciriptase enzymes converts RNA encoded genes into DNA and this DNA is transported into the nucleus and integrated into cells chromosomal DNA.

It is programmed that the packaging cell line to produce the essential structural viral genes that have been deleted from the viral genome. Since the viral genes are encoded in the packaging cells chromosome, the resulting viral particles are incapable of causing disease, however, they act as carriers for the desired, genes of therauptic nature. The retro virus vector thus resulted, can be used only with dividing cells because cells must undergo mitosis before gene integration takes place. However this feature limits in vivo use in cases like cancer supression but is not a limitation on ex-vivo systems.

A bioprocess limitation is another one for retrovirus system, for instance, the production of high titer virus and subsequent purification and concentration without loss of infectivity is difficult. The average number of genes delivered to target cells is too low to achieve a beneficial effect is another limitation to the effectiveness of many gene therapy approaches, including the use of retroviruses. The low efficiency is attributed to the bioprocess limitation on production of a concentrated, highly infections retrovirus preparation.

Two factors are responsible problems in producing a highly infectious preparation

1. Rapid decay of virus and 2. inhibition of viral infectivity by proteoglycans released by the packaging cell lime. The molecular weight of proteoglycans is very high when virus is concentrated by cultra filtration, the proteglycan will also be concentrated by the same factor.

The number of cells infected does not increase significantly eventhough number of injected increase with high virus concentration due to increased inhibition of infection by proteoglycans and in some case infected cells may decrease with high viral concentrations. Producing virus at reduced temperatures i.e., 28 $^{\circ}$C may be another approach to increase viral titre. Rate of virus production and viral decay decrease with temperature but the rate of decay is more sensitive to temperature. The amount of potentially active virus is higher when produced at 28 °C.

Stem Cell

Bioreactors may be used for mass production of cells for transplantation. Extensive replication and self-renewal is expected in some animal cells. Other cells are highly differentiated and perform specific functions, normally differentiated cells cannot replicate. So a stem cell is an undifferentiated cell capable of continuous self-renewal, which can also produce more numbers of differentiated progeny, depending on the extracelluar factors present. Hematopoietic system may be studied here. Human circulating system consists of eight major types of fully mature blood cells. Two types of progenitor cells are produced by a hematopoietic stem cells.

The replication of progenitor cell is expected but the progenitor cells have more restricted range of cells in which they may differentiate. Myclorid cells i.e., red blood cells or erythrocytes, platelets, macrophages etc; are produced by one type of progenitor cell lymphoid cells T.cells, B.cells are produced by another type of progenitor cells in the hematopoietic system. A large number of hematopoietic growth factors are responsible for the progression of differentiation. The process of generating blood cells namely Hematopieses occurs in bone marrow. Bone marrow transplants may be supplemented by ex. Vivo hematopoieses.

Commercial scale systems for hematopoiesis may be developed. Bioreactions must accomodate adherent cell growth since coculture with stromacells from the bone marrow is necessary to generate required growth factors. Mini-structures which are much like that in bone may be incorporated in fluidized bed reactors. It may be difficult to have mix type of cells like in bone. Flat bed reactors may be used for this purpose. Also membrane based units may be used.

References

1. Bronzinno J.D. (2000), "The Biomedical Engineering Hand book" CR c press LLC, BocaRation F L.
2. Le Doux, J.M, H.E, Davis, J.R. Morgan, and M.L. Yarmush, (1999.). "Kinetics of Retrovirus Production and Decay", Biotechnol. Bioeng.
3. Naughton, G.K.,Skin. (1999.)", The First Tissue Engineered Products, The advanced Tissue sciences story sci. Am.

Review Questions

1. Write about bioprocess engineering applied per medical applications.
2. Explain tissue culture process.
3. Describe culturing of chrodocycle.
4. Write about Gene Thearpy.
5. Explain about stem cell.

Multiple Choice Questions

Chapter - 1

1. Penicillin is discovered by　　　　　　　　　　　　　　　　[　　]
   (a)　heknel　　　　　　　　　　　(b)　lousis Pasteur
   (c)　monod　　　　　　　　　　　(d)　alexander Fleming

2. Penicillin chrysogenum produces　　　　　　　　　　　　[　　]
   (a)　streptomycin　　　　　　　　(b)　penicillin
   (c)　neomycin　　　　　　　　　　(d)　tetracycline

3. Vitamin B12 is produced by　　　　　　　　　　　　　　[　　]
   (a)　chemical synthesis　　　　　　(b)　microbial conversion
   (c)　hydrolysis　　　　　　　　　　(d)　reduction

4. In solid state fermentation the reactor contents are　　　　[　　]
   (a)　mixed　　　　　　　　　　　　(b)　not mixed
   (c)　agitated　　　　　　　　　　　(d)　kept stationary

5. Microbial cultures are stored at low temperatures　　　　　[　　]
   (a)　$< 5^0C$　　　　　　　　　　　(b)　$< 0^0C$
   (c)　$>10^0C$　　　　　　　　　　　(d)　$>15^0C$

6. Subculture is developed from　　　　　　　　　　　　　　[　　]
   (a)　inoculum　　　　　　　　　　(b)　soil culture
   (c)　stock culture　　　　　　　　　(d)　air culture

7. Inoculum is developed from　　　　　　　　　　　　　　[　　]
   (a)　stock culture　　　　　　　　　(b)　from soil
   (c)　from water　　　　　　　　　　(d)　subculture

8. Inoculum will be % of reactor volume　　　　　　　　　　[　　]
   (a)　5%　　　　　　　　　　　　　(b)　2%
   (c)　15%　　　　　　　　　　　　　(d)　50%

9.   Growth time required for lyophilized cultures is   [   ]
      (a)   1-5 days   (b)   3-14 days
      (c)   15-20 days   (d)   20-30 days

10.   Growth time for frozen culture for bacteria is   [   ]
      (a)   1-2 hrs   (b)   50-40 hrs
      (c)   4-48 hrs   (d)   72-96 hrs

11.   Growth time for frozen culture for fungi   [   ]
      (a)   1- 7 days   (b)   10-15 days
      (c)   15-20 days   (d)   8-10 days

12.   In large bioreactors the height to diameter ratio is   [   ]
      (a)   2.0   (b)   1.0
      (c)   3.0   (d)   5.0

13.   Product from fermentation broth is isolated by   [   ]
      (a)   evaporation   (b)   leaching
      (c)   precipitation   (d)   diffusion

14.   Alcoholic fermentation is   [   ]
      (a)   aerobic reaction   (b)   anaerobic reaction
      (c)   chemical reaction   (d)   noncatalytic reaction

15.   Commercialisation of penicillin was done by   [   ]
      (a)   merk, USA   (b)   squibb, USA
      (c)   pfizer, USA   (d)   dow chemicals, USA

Chapter - 2

16.   Microbiology is the study of   [   ]
      (a)   fungi   (b)   molds
      (c)   bacteria   (d)   microorganisms

17.   Prokaryotes do not have well defined   [   ]
      (a)   cell membrane   (b)   cytoplasm
      (c)   nuclear membrane   (d)   endoplasm

18. The volume of prokaryote cells is about   [   ]
    (a) 10^{-20} ml per cell   (b) 10^{-30} ml per cell
    (c) 10^{-12} ml per cell   (d) 10^{-25} ml per cell

19. The mass of single prokaryotic cell is   [   ]
    (a) 10^{-30} g   (b) 10^{-40} g
    (c) 10^{-10} g   (d) 10^{-12} g

20. The doubling time of prokaryotes is about   [   ]
    (a)   5 min   (b)   20min
    (c)   30 min   (d)   60 min

21. The cell wall thickness of prokaryotes is about   [   ]
    (a)   50 A^0 (b)   100 A^0
    (c)   200 A^0 (d)   300 A^0

22. The example for prokaryotes carrying out photosynthesis   [   ]
    (a)   fungi   (b)   blue green algae
    (c)   amoeba   (d)   bacteria

23. In eukaryotes the catalytic activity at ribosomes is controlled   [   ]
    (a)   mitochondria   (b)   nucleus
    (c)   golgibodies   (d)   lysosomes

24. The size of the eukaryotic cell is   [   ]
    (a)   1µm   (b)   2 µm
    (c)   3 µm   (d)   >> 5 µm

25. The kingdom protista in 1866 was proposed by   [   ]
    (a)   fleming   (b)   henri
    (c)   haekel   (d)   monod

26. Bacteria that retains blue crystal violet colour after Gram staining is called   [   ]
    (a)   gram positive   (b)   gram negative
    (c)   stained   (d)   unstained

27. The length of yeast varies from   [   ]
    (a)   1-2µm   (b)   5-30 µm
    (c)   30-40 µm   (d)   50-60 µm

28. The common reproduction in yeast is done by   [   ]
    (a)   asexual   (b)   sexual
    (c)   budding   (d)   fragmentation

29. Citric acid is produced by   [   ]

    (a) bacteria   (b) yeast

    (c) A.niger   (d) mold

30. Polymers of nucleotides form into   [   ]

    (a) amino acids   (b) nucleic acids

    (c) organic acids   (d) inorganic acids

31. Nucleotides mainly contain   [   ]

    (a) one component   (b) two components

    (c) three components   (d) four components

32. Nucleotides consists of   [   ]

    (a) nucleoside and phosphate   (b) nucleoside and sulphate

    (c) nucleoside and nitrogen   (d) nucleoside and oxygen

33. Purines contain two nitrogen bases that are   [   ]

    (a) adenine and thymine   (b) adenine and guanine

    (c) adenine and cytosine   (d) adenine and uracil

34. Energy obtained from nutrients is normally stored as   [   ]

    (a) ADP   (b) NAD

    (c) ATP   (d) NADH

35. In DNA and RNA nucleotides are connected between   [   ]

    (a) 1st and 2nd carbon   (b) 2nd and 4th carbon

    (c) 3rd and 5th carbon   (d) 1st and 3rd carbon

36. Ribosomes contain three different rRNAs   [   ]

    (a) 23S ,16S and 5S   (b) 20S , 15 S, and 3S

    (c) 20S, 16S and 3S   (d) 23S, 15S and 5S

37. Proteins are made up of four elements   [   ]

    (a) carbon, hydrogen, sulphur and oxygen

    (b) carbon, hydrogen, nitrogen and oxygen

    (c) carbon, hydrogen, phosphorous and oxygen

    (d) carbon, hydrogen, phosphorous and sulphur

38. Aminoacids are recognized as per the R group attached to   [   ]

    (a) α-carbon   (b) β -carbon

    (c) γ-carbon   (d) no carbon

39. Total hydrolysis of proteins is carried out by heating them   [   ]

    (a) 1 N HCl   (b) 6 N HCl

    (c) 3 N HCl   (d) 5 N HCl

40. At low pH the aminoacid is   [   ]

    (a) positively charged   (b) negatively charged

    (c) neutral charge   (d) dipolar

Chapter - 3

41. The standard free energy for ATP is about   [   ]

    (a) 5.0 kcal/mol   (b) 7.3 kcal/mol

    (c) 8.0 kcal/mol   (d) 10.0 kcal/mol

42. In EMP pathway the important metabolite of the metabolism is   [   ]

    (a) pyruvate   (b) lactate

    (c) fructose   (d) glucose

43. In TCA cycle pyruvate is converted to   [   ]

    (a) CO_2 and NAD   (b) CO_2 and NADH

    (c) CO_2 and ATP   (d) CO_2 and ADP

44. The sequence in respiratory chain is commonly known as   [   ]

    (a) EMP chain   (b) electron transport chain

    (c) phosphorylation   (d) oxidation

45. In electron transport chain the energy deposited in 36 mol of ATP is   [   ]

    (a) 250 kcal/mol glucose   (b) 263 kcal/mol glucose

    (c) 200 kcal/mol glucose   (d) 260 kcal/mol glucose

46. When ATP/ADP ratio is high, the enzyme is   [   ]

    (a) activated   (b) inactivated

    (c) reduced activity   (d) increased activity

47. The rate of glycolysis under anaerobic condition will be   [   ]

    (a) equal to aerobic condition   (b) higher than aerobic condition

    (c) less than aerobic condition (d) lower than aerobic condition

48. High ratios of NADH/NAD$^+$ results in [   ]
   (a) low rate of glycolysis   (b) high rate of glycolysis
   (c) no change in glycolysis   (d) equal rate of glycolysis

49. Some steps in electron transport chain could be inhibited by [   ]
   (a) cyanide   (b) cyanide aze
   (c) phenol   (d) benzene

50. The first step in metabolism of hydrocarbons is [   ]
   (a) reduction   (b) hydrolysis
   (c) oxygenation   (d) hydrogenation

51. The important intermediate during oxidation of aromatic hydrocarbons in metabolism is [   ]
   (a) phenol   (b) catechol
   (c) pyruvic acid   (d) benzene

52. This is used in biosynthesis [   ]
   (a) NAD$^+$ (b) NADH
   (c) NADPH   (d) ATP

53. In pentose phosphate path way gleceraldehyde 3 phosphate is converted to [   ]
   (a) pyruvate   (b) lactic acid
   (c) benzoic acid   (d) carboxylic acid

54. The production of glucose is known as, [   ]
   (a) synthesis   (b) gluconogenesis
   (c) biosynthesis   (d) genesis

55. In glycolysis the step of conversion of phosphophenol into pyruvate is [   ]
   (a) reversible   (b) irreversible
   (c) forward reaction   (d) backward reaction

Chapter - 4

56. Specific growth rate is expressed as [   ]
   (a) $\mu_{net} X = dx/dt$ (b) $\mu_{net} X = -dx/dt$
   (c) $\mu_{net} = -1/X\ dx/dt$ (d) $\mu_{net} X = 1/X\ dx/dt$

57. The inoculum culture age plays an important role on [   ]
   (a) log phase   (b) lag phase
   (c) stationary phase   (d) decay pahse

58. The doubling time of microorganism is given as t_d [   ]

    (a) $0.652/\mu_{net}$ (b) $0.683/\mu_{net}$

    (c) $0.693/\mu_{net}$ (d) $0.663/\mu_{net}$

59. The conversion of cell mass into maintenance energy during stationary phase is given by (hyy   [   ]

    (a) $dx/dt = k_d\,X$ (b) $dx/dt = -\,k_d\,X$

    (c) $dx/dt = k_d\,1/X$ (d) $dx/dt = -1/X\,k_d$

60. Ethanol is produced by fermentation of sugar by fermentation of sugar   [   ]

    (a)   Bacillus subtilis   (b)   Escherisia coli

    (c)   Saccharomysis cervisiae   (d)   Trichoderma viridie

61. The microbial decay rate is expressed as   [   ]

    (a) $dN/dt = -\,k_d\,N$ (b) $dN/dt = k_d\,N$

    (c) $dN/dt = k_d\,1/N$ (d) $dN/dt = -1/N\,k_d$

62. Yield coefficient is defined by   [   ]

    (a) $Y_{X/S} = \Delta X/\Delta S$ (b) $Y_{X/S} = -\Delta X/\Delta S$

    (c) $Y_{X/S} = \Delta S/\Delta X$ (d) $Y_{X/S} = -\Delta S/\Delta X$

63. The specific growth differ from μ_{net} when undergoes metabolism is   [   ]

    (a) 0 (b) > 0

    (c) -2 (d) -1

64. The specific growth rate of product formation is given by   [   ]

    (a) $q_p = \alpha\,\mu_g - \beta$ (b) $q_p = \alpha\,\mu_g + \beta$

    (c) $q_p = \beta - \alpha\,\mu_g$ (d) $q_p = 1/\alpha\,\mu_g + -\beta$

65. Monod equation for growth kinetics is   [   ]

    (a) $\mu_g = -\,\mu_m\,S/K_s + S$ (b) $\mu_g = \mu_m\,S/K_s + S$

    (c) $\mu_g = \mu_m\,S/K_s - S$ (d) $\mu_g = 1/S\,\mu_m\,/K_s + S$

66. When endogeneous metabolism is neglected   [   ]

    (a) $\mu_g = -\,\mu_{net}$ (b) $\mu_g = S\,\mu_{net}$

    (c) $\mu_g = \mu_{net}$ (d) $\mu_g = 1/S\,\mu_{net}$

67. The equation for cell death condition is given by   [   ]

    (a) $\mu_g = \mu_m\,S/K_s + S + k'_d$ (b) $\mu_g = 1/S\,\mu_m/K_s + S - k'_d$

    (c) $\mu_g = \mu_m\,S/K_s + S - k'_d$ (d) $\mu_g = S\,\mu_m\,/K_s + 1 - k'_d$

68. Secondary metabolites are products which are   [   ]

(a) growth associated   (b) non growth associated

(c) depends on substrate   (d) depends on product

69. During lad phase the microbial cells synthesize new enzymes suiting to   [   ]

(a) previous medium   (b) new medium

(c) nutrients   (d) products

70. During lag phase cell mass may increase but the number of cells   [   ]

(a) may increase   (b) may decrease

(c) remains same   (d) instantaneous reduction

Chapter 5

71. The physical process of immobilization of enzymes is known as   [   ]

(a) entrainment   (b) entrapment

(c) weak binding   (d) strong binding

72. The advantage of matrix entrapment is that enzymes is   [   ]

(a) chemically modified   (b) not denatured

(c) not chemically modified   (d) remains stable

73. During entrapment enzyme leakage can be avoided by   [   ]

(a) increasing MW cut off   (b) same MW cut off

(c) reducing MW cut off   (d) altering MW cut off

74. In adsorption the enzymes are physically adsorbed by   [   ]

(a) ionic forces   (b) Vander waals forces

(c) inertial forces   (d) chemical forces

75. In covalent binding the functional groups in active site must be   [   ]

(a) present in low number   (b) present in more number

(c) must not be present   (d) must be present in excess

76. The another method of enzyme immobilization is the cross linking of enzyme molecules with agents such as   [   ]

(a) 2,2-disulfonic acid   (b) 1,2-disulfonic acid

(c) dilute sulfuric acid   (d) dilute hydrochloric acid

77. In covalent binding , the binding capacity is a function of   [   ]

   (a)   area of surface   (b)   hydrophobicity of surface

   (c)   hydrophilicity of surface   (d)   density of support surface

78. In CSTRs the only disadvantage would be disruption of enzyme pellets are due to   [   ]

   (a)   stress   (b)   inertial force

   (c)   shear force   (d)   gravitational force

79. Fluidised bioreactors exhibit good features of   [   ]

   (a)   CSTR + packed bed bioreactor   (b)   CSTR + loop bioreactor

   (c)   CSTR+ air lift reactor   (d)   CSTR + bubble column reactor

80. The advantage of rDNA technology is useful in producing rare enzymes   [   ]

   (a)   at good quality   (b)   at high cost

   (c)   in large quantity at low cost   (d)   in large quantity at high cost

81. Out of the total enzymes demand , the share of protease is   [   ]

   (a)   <30%   (b)   > 60%

   (c)   <10%   (d)   <5%

82. The enzymes used in detergents for removal of stains are   [   ]

   (a)   amylases   (b)   proteases

   (c)   pectinases   (d)   isomerases

83. Lipases are produced from   [   ]

   (a)   bacteria   (b)   fungi

   (c)   animal pancreas   (d)   plant cells

84. Glucoamylase hydrolyses amylopectin portion of starch and is known as   [   ]

   (a)   hydrolyzing enzyme   (b)   saccharifying enzyme

   (c)   liquefying enzyme   (d)   tenderizing enzyme

85. The enzymes used in production of semisynthetic penicillins are   [   ]

   (a)   amylase   (b)   pectinase

   (c)   penicillin acylase   (d)   penicillin amylase

86. Meat tenderization is carried out by   [   ]

   (a)   invertase   (b)   papain

   (c)   amylase   (d)   protease

87. Isofunctional enzymes may be used for   [   ]

    (a)   inhibition of pathways   (b)   repression of pathways

    (c)   promotion of pathways   (d)   diverting the pathways

88. In passive diffusion molecules move   [   ]

    (a)   from low to high concentration   (b)   from high to low concentration

    (c)   due to flux   (d)   dueto intracellular concentration

89. Passive diffusion facilitates cellular uptake of   [   ]

    (a)   water and oxygen   (b)   water and hydrogen

    (c)   water and nitrogen   (d)   water and carbondioxide

90. In eukaryotic cells sugars and other low molecular weight compounds are diffused due to   [   ]

    (a)   passive diffusion   (b)   transport diffusion

    (c)   active transport   (d)   cellular transport

Chapter 6

91. The molecular weight of proteins vary from   [   ]

    (a)   1000 to 5000 daltons   (b)   5000 to 10,000 daltons

    (c)   15000 to several million Daltons   (d)   10,000 to 15000 daltons

92. The enzymes that contain non protein group is called   [   ]

    (a)   apoenzyme   (b)   holoenzyme

    (c)   cofactor   (d)   coenzyme

93. The formation of enzyme-substrate complex directs   [   ]

    (a)   enhancing energy   (b)   lowering of activation energy

    (c)   neglecting activation energy   (d)   controls activation energy

94. In proximity effect enzymes hold   [   ]

    (a)   multiple substrates close to each other

    (b)   multiple products close to each other

    (c)   complex molecules close to each other

    (d)   simple molecules close to each other

95. enzymes catalysed kinetics was developed by   [   ]

    (a)   Monod   (b)   Michalis –Menten

    (c)   Crick-watson   (d)   Louis Pasteur

96. Quasi steady state approximation was first proposed by　　　　[　　]
    (a) Brigs-Haldane　　　　(b) Michalis-Menten
    (c) Monod　　　　(d) Crick –Watson

97. Michalis-Menten parameters like Km may be estimated by　　　　[　　]
    (a) Briggs –Haldane　　　　(b) Crick-Watson
    (c) Line weaver Burk plot　　　　(d) monod

98. In uncompetitive inhibition the inhibitors bind　　　　[　　]
    (a) enzymes　　　　(b) substrates
    (c) products　　　　(d) ES-complex

99. The enzyme substrate complex is formed due to weak forces like　　　　[　　]
    (a) ionic forces　　　　(b) inertial forces
    (c) Vander waals forces　　　　(d) attraction forces

100. Some enzymes for their proper functioning require　　　　[　　]
    (a) coenzymes　　　　(b) cofactors
    (c) coenzymes-cofactors　　　　(d) apoenzymes

101. The activation energy for enzyme catalysed reactions at $20^0 C$ would be　　[　　]
    (a) 20 kcal/mol　　　　(b) 15 kcal/mol
    (c) 7 kcal/mol　　　　(d) 10 kcal/mol

102. In rapid equilibrium assumption , the equilibrium coefficient could be used to express (ES) in terms of　　　　[　　]
    (a) (P)　　　　(b) (E_0)
    (c) (S)　　　　(d) (E)

103. In competitive inhibition K'_{map} is increased and reaction rate is　　　　[　　]
    (a) increased　　　　(b) reduced
    (c) not altered　　　　(d) no effect

104. Noncompetetive inhibitors bind enzyme sites which are　　　　[　　]
    (a) not active　　　　(b) active
    (c) not available　　　　(d) available

105. In substrate inhibition the inhibition effect is not observed at　　　　[　　]
    (a) high substrate concentration　　(b) low substrate concentration
    (c) low enzyme concentration　　(d) high enzyme concentration

Chapter 7

106. In batch bioreactors all the reactants are fed to the bioreactor   [   ]

    (a)   in the middle of the reaction

    (b)   initially during start up of the reaction

    (c)   initially and during middle of the reaction

    (d)   initially and during end of the reaction

107. In continuous bioreactors inflow and out flow will be continuous and the product can be drawn   [   ]

    (a)   intermittently   (b)   at the middle of the reaction

    (c)   continuously   (d)   at the end of the reaction

108. In fed batch reactors nutrients and substrates are fed continuously and product is removed   [   ]

    (a)   at the middle of the batch   (b)   at the end of the batch

    (c)   at the beginning of the batch   (d)   intermittently during the batch

109. In air lift reactors mixing and internal circulation is achieved by   [   ]

    (a)   recirculation   (b)   by passing air

    (c)   by removing air   (d)   intermittent circulation

110. In bubble column reactor the aspect ratio is   [   ]

    (a)   high   (b)   low

    (c)   average   (d)   maximum

111. In fluidized bed bioreactors the gravitational force acting down on the particle is counter acted by   [   ]

    (a)   inertial forces   (b)   bouancy forces

    (c)   centrifugal forces   (d)   frictional forces

112. The dynamic characteristics of a bioreactor are a function of   [   ]

    (a)   concentration changes of product

    (b)   concentration changes of enzyme

    (c)   concentration changes of fci

    (d)   concentrationchanges of intermediates

113. In chemostat model the steady state condition depends on   [   ]

    (a)   feed rate   (b)   dilution rate

    (c)   product rate   (d)   enzyme rate

114. In Monod chemostat wah out occur when μ_m is   [   ]

    (a)   equal to 'd"   (b)   less than "D"

    (c)   more than "D"   (d)   equal to zero

115. In enzyme catalysed reactions the equation for substrate inhibition is   [   ]

    (a)   Vmax/1+km/s+S/ki   (b)   Vmax/1+Km/s –S/ki

    (c)   Vmax/1-km/s +s/ki   (d)   Vmax/1-km/s-s/ki

116. The CSTR with separator with outlet stream and recycle from separator to feed will   [   ]

    (a)   reduce biomass in reactor   (b)   enhance biomass in reactor

    (c)   no chage biomass in reactor   (d)   alters biomass in reactor

117. In CSTR recycle production rate is greater than the nonrecycle production rate by   [   ]

    (a) $[\,1\text{-}a(b\text{-}1)]^{-1}$ (b) $[\,1\text{-}a(b\text{+}1)]^{-1}$

    (c) $[\,1\text{+}a(b\text{-}1)]^{-1}$ (d) $[\,1\text{+}a(b\text{+}1)]^{-1}$

118. If the initial feed to the batch reactor and feed at entry point of plug flow reactor are same, the residence time of PFR is   [   ]

    (a)   more   (b)   same

    (c)   less   (d)   zero

119. Exit age distribution is known as the study of   [   ]

    (a)   all the fractions of traces   (b)   some fractions of traces

    (c)   no fraction of traces   (d)   a few fractions of traces

120. The conversion of the reactant can be estimated by measuring   [   ]

    (a)   concentration of tracer   (b)   exit age distribution of the tracer

    (c)   volume of tracer   (d)   density of tracer

121. Dispersion model depends on certain assumptions like   [   ]

    (a)   large short circuiting   (b)   no short circuiting

    (c)   a few short circuiting   (d)   less short circuiting

122. Mass transfer coefficient Kla depends on   [   ]

    (a)   bubble size   (b)   bubble volume

    (c)   bubble density   (d)   bubble area

Chapter 8

123.   The destruction of microorganisms follows   [   ]

    (a)   zero order reaction   (b)   first order reaction

    (c)   second order reaction   (d)   nth order reaction

124.   Bacterial cells are more resistance to heat than   [   ]

    (a)   plant cells   (b)   animal cells

    (c)   vegetative cells   (d)   matured cells

125.   While determining the life span of spores probabilistic approach is used and for media sterilization "N" is,   [   ]

126.   The probability (Pt) the life spore n total is < t at a given temperature is given as   [   ]

    (a) $P(t) = (1 - e^{-kt})^n$ (b) $P(t) = (1 + e^{-kt})^n$

    (c) $P(t) = (1 - e^{kt})^n$ (d) $P(t) = (1 + e^{kt})^n$

127.   The plate type of sterilization is commonly used in   [   ]

    (a)   chemical industry   (b)   textile industry

    (c)   fermentation industry   (d)   inorganic chemical industry

128.   The size of the dust particles in air sterilization varies from   [   ]

    (a)   0.5 to 1.0 μ   (b)   1.0 to 2.0 μ

    (c)   2.0 to 3.0 μ   (d)   4.0 to 5.0 μ

129.   Air sterilization can be achieved by   [   ]

    (a)   X-rays   (b)   γ-rays

    (c)   U-V rays   (d)   Sun rays

130.   The germicidal spray used in air conditioning equipment is   [   ]

    (a)   hydrogen peroxide   (b)   ethylene oxide

    (c)   nitrous oxide   (d)   carbon dioxide

131.   The efficiency of the filter may be defined as ratio of number of particles   [   ]

    (a)   final number of particles   (b)   initial number of particles

    (c)   total number of particles   (d)   all particles of various sizes

132. X_{90} may be defined as the   [   ]

    (a)   depth of filter to remove 90% of particles

    (b)   width of filter to remove 90% of particles

    (c)   thickness of filter to remove 90% of particles

    (d)   diameter of filter to remove 905 of particles

133. The temperature used for short time for destroying spores is   [   ]
    (a) 100^0C (b) 50^0C
    (c) 150^0C (d) 200^0C

134. The steam sterilisation conditions for bioreactors are   [   ]
    (a) 12 psig for 30 min   (b) 15 psig for 20 min
    (c) 15 psig for 30 min   (d) 12 psig for 20 min

135. Del factor may be defined as measure of fractional reduction in   [   ]
    (a) living organism count   (b) viable organism count
    (c) dead organism count   (d) nonviable organism count

136. One disadvantage of batch sterilization is   [   ]
    (a) destruction of substrates   (b) destruction of products
    (c) destruction of nutrients   (d) destruction of enzymes

137. Design equation for calculating temperature time profile for electrical heating is   [   ]
    (a) $T = T_0(1-\alpha t)$ (b) $T = T_0(1+\alpha t)$
    (c) $T = T_0(1+\alpha t/T_0)$ (d) $T = T_0(1/T_{0+}\alpha t)$

Chapter 9

138. Penicillin was discovered by   [   ]
    (a) Pasteur   (b) Alexander Fleming
    (c) Henry   (d) Briggs

139. The vegetative cells may be suspended in glycerol and stored at   [   ]
    (a) -20^0C (b) -50^0C
    (c) -60^0C (d) -70^0C

140. The production temperature of penicillin is   [   ]
    (a) $30-32^0C$ (b) $25-28^0C$
    (c) $20-25^0C$ (d) $10-15^0C$

141. Citric acid is produced by using   [   ]
    (a) Aspergillus niger   (b) B.subtilis
    (c) Pseudomonas sp.   (d) E.coli

142. Bakers yeast is produced by   [   ]
    (a) anaerobic fermentation   (b) aerobic fermentation
    (c) anoxic fermentation   (d) aerobic-anaerobic fermentation

143. Ethyl alcohol is produced by   [   ]
     (a)   E.coli   (b)   S.cervisae
     (c)   B.subtilis   (d)   A.niger

144. Protective antigen can be produced by using genetically engineered   [   ]
     (a)   Bacillus subtilis   (b)   Aspergillus niger
     (c)   Pseudomonas species   (d)   Trichoderme viridie

145. Enzymes are basically   [   ]
     (a)   organic molecules   (b)   proteins
     (c)   cofactors   (d)   holozyme

146. HFCS containing 55% of fructose is used in   [   ]
     (a)   food products   (b)   jams
     (c)   jellies   (d)   soft drinks

147. The airlift fermentor contains a draft tube installed   [   ]
     (a)   centre of fermentor   (b)   top of thr fermentor
     (c)   bottom of fermentor   (d)   side of the fermentor

148. Power input to a fermentor of pilot plant is   [   ]
     (a)   6-8 W/l   (b)   8-10 w/l
     (c)   3-5 w/l   (d)   10-15 W/l

149. Solid state fermentation largely in commercial production of   [   ]
     (a)   proteins   (b)   enzymes
     (c)   organic acids   (d)   vaccines

150. Submerged fermentation has advantage over solid state fermentation due to [   ]
     (a)   better process control   (b)   better production
     (c)   better yields   (d)   better conversion

151. Material of construction often used for fermentors is   [   ]
     (a)   glass or stainless steel   (b)   mild steel
     (c)   cast iron   (d)   aluminium

152. Corn starch is gelatinized to produce dextrose at   [   ]
     (a) 40^0C (b) 50^0C
     (c) 65^0C (d) 70^0C

153. Cholesterol oxidase is mostly used in   [   ]
     (a)   medicine   (b)   analytical purpose
     (c)   textiles   (d)   detergent industries

Chapter 10

154. The resistances used during oxygen transfer from gas to liquid are [   ]

    (a) 5 (b) 4

    (c) 7 (d) 8

155. The approximate consumption of oxygen by yeast during respiration

    would be [   ]

    (a) 0.5 g oxygen/g of dry cell (b) 0.8 g oxygen/g of dry cell

    (c) 0.3 g oxygen/g of dry cell (d) 1.0 g oxygen/g of dry cell

156. The size of the single bubble d_B is proportional to [   ]

    (a) square root of orfice dia (b) cube root of orfice dia

    (c) squqare of orfice dia (d) cube of orfice dia

157. Power number concept was developed by [   ]

    (a) Rushton (b) Baily

    (c) Newton (d) Haldane

158. Power number is the ratio of [   ]

    (a) inertial force/external force

    (b) External force/inertial force

    (c) gravitational force/inertial force

    (d) external force/gravitational force

159. Modified Reynolds number is the ratio of [   ]

    (a) inertial force/viscous force (b) external force/viscous force

    (c) viscous force/inertial force (d) viscous force/inertial force

160. The degree of power decrease in gassed to ungassed system varies from [   ]

    (a) 0.8 to 1.0 (b) 0.7 to 1.0

    (c) 0.9 to 1.0 (d) 0.3 to 1.0

161. The aeration number 'Na' would be useful in evaluating [   ]

    (a) solubility (b) degree of dispersion

    (c) mass transfer (d) resistance

162. For each type of impeller the value of Np at high values of N_{Re} [   ]

    (a) increases (b) decreases

    (c) remains constant (d) varies

163.   The oxygen consumption at peak for a population density of 10^9 cells/ml is estimated by considering cell volume to be about   [   ]

    (a) 10^{-5} ml   (b) 10^{-9} ml

    (c) 10^{-7} ml   (d) 10^{-10} ml

164.   The oxygen consumption rate of actively respiring microorganisms is more than the oxygen saturation value per hour by   [   ]

    (a)   700 times   (b)   500 times

    (c)   100 times   (d)   50 times

165.   In gas liquid operations the film transfer coefficient Kl can be estimated by equation   [   ]

166.   While designing fermentors if volumetric coefficient of oxygen Kca or Kv are not available , a plot is used between,   [   ]

    (a)   Absorption number Vs power number

    (b)   Absorption number Vs aeration number

    (c)   Absorption number Vs power to unit volume

    (d)   Absorption number Vs Reynolds number

167.   The critical oxygen values for organisms vary in the range of   [   ]

    (a)   0.001 to 0.005 mmol/l   (b)   0.003 to 0.008 mmol/l

    (c)   0.003 to 0.05 mmol/l   (d)   0.003 to 0.02 mmol/l

168.   During bubble aeration of water Kla values decreased when 1% peptone was added , to about,   [   ]

    (a)   ½ of initial value   (b)   1/3 rd of initial value

    (c)   ¼ th of initial value   (d)   1/5 th of initial value

Chapter 11

169.   The costs of separations during recovery of bio products are more than manufacturing costs by   [   ]

    (a)   20%   (b)   30%

    (c)   50%   (d)   40%

170.   The cells from fermentation broth are removed by using centrifugation varies between   [   ]

    (a)   50 to 0.1 μm   (b)   60 to 0.1 μm

    (c)   80 to 0.1 μm   (d)   100 to 0.1 μm

171. For discs or bowl centrifuge scale up factor depends on   [   ]

    (a) one radial distance   (b) two radial distances

    (c) three radial distances   (d) four radial distances

172. Penicillin mycelia are removed by using   [   ]

    (a) rotary vacuum filter   (b) membrane separator

    (c) reverse osmosis   (d) plate and frame filter

173. The drum of vacuum filter is covered with   [   ]

    (a) cloth   (b) pre coat

    (c) with alum   (d) a polymer

174. The resistance of filter medium r_m depends on characteristics of   [   ]

    (a) filter medium   (b) filter cloth

    (c) filter drum   (d) vacuum

175. In ultrasonic vibrations the wave density used is   [   ]

    (a) 10 kc/s   (b) 15 kc/s

    (c) 20 kc/s   (d) 30 kc/s

176. In French press the cell paste is filled in hollow cylinder in a stainless steel Block and pressure applied is   [   ]

    (a) low   (b) medium

    (c) average   (d) high

177. Some mechanical homogenisers contain small beads of   [   ]

    (a) 10-15 mesh size   (b) 1-5 mesh size

    (c) 20-50 mesh size   (d) 15-20 mesh size

178. For cell disruption in X-press the pressure levels applied is   [   ]

    (a) 100 bar   (b) several hundred bars

    (c) 50 bar   (d) 20 bar

179. Bacteria cell wall may be lysed by using enzymes like   [   ]

    (a) amylase   (b) lysozyme

    (c) protease   (d) catalase

180. The inhibitory fermentation products such as ethanol are extracted by   [   ]

    (a) distillation   (b) membrane separation

    (c) liquid extraction   (d) evaporation

181. Podbielneak extractors are commonly used for extracting   [   ]

    (a)   ethanol   (b)   antibiotics

    (c)   organic acids   (d)   enzymes

182. In two phase affinity partition extraction kp values may be increased by adding   [   ]

    (a)   PEG-ATP   (b)   PEG-NAD

    (c)   PEG- NADH   (d)   PEG-ADP

183. Precipitation of protein by reducing dielectric constant of solution occurs by adding organic solvents at   [   ]

    (a) $T < 0^0C$ (b) $T < -2^0C$

    (c) $T < -5^0C$ (d) $T < 2^0C$

184. Macromolecules are separated by ultra filters with a molecular weight range of   [   ]

    (a)   1000-2000   (b)   100-1000

    (c)   1000-1500   (d)   2000- 5, 00,000

Chapter 12

185. During scale up tip velocity of impeller, the F/V value   [   ]

    (a)   increases   (b)   decreases

    (c)   remains same   (d)   varies

186. The power requirements of an impeller in gassed systems can be represented in geometrically similar systems by   [   ]

    (a)   P   (b)   V

    (c)   F/V   (d)   V/F

187. Impellers used are flat blade turbine type when Dt/Di are ranged from   [   ]

    (a)   1.0 to 2.0   (b)   2.02 to 3.4

    (c)   3.0 to 4.0   (d)   4.0 to 5.0

188. Bartholomew for finding out oxygen transfer coefficient in fermentation applied   [   ]

    (a)   cromite oxidation   (b)   sulfite oxidation

    (c)   nitrite oxidation   (d)   bromite oxidation

189. While scaling down heterogenicity in temperature may   [   ]

    (a)   be reproduced   (b)   not be reproduced

    (c)   be varied   (d)   be difficult to reproduce

190. Equation for OUR can be written as   [   ]

    (a) $OUR = 1/Kla(C^*-Cl)$ (b) $OUR = Kla (C^*-Cl)$

    (c) $OUR = 1/Kla =(C^*+Cl)$ (d) $OUR = Kla (C^*-Cl)$

191. The respiration rate of E.coli in mmol O_2 /g dw-in is   [   ]

    (a)   10-12   (b)   20-30

    (c)   0-10   (d)   15-20

192. Antibiotics yields are better at power to volume ratio in Hp/m3 is   [   ]

    (a)   0-1   (b)   0.5-1.0

    (c)   > 1.5   (d)   1.0- 1.5

193. The specific rate of ethanol production depends on   [   ]

    (a)   height of bioreactor   (b)   type of bioreactor

    (c)   volume of bioreactor   (d)   area of bioreactor

194. Impeller tip velocity is   [   ]

    (a) $v \alpha 1/ni\ Di$ (b) $v \alpha ni/ Di$

    (c) $v \alpha ni\ Di$ (d) $v \alpha Di/ni$

195. The power is related in agitated bioreactors by   [   ]

    (a) $P \alpha 1/n^3 Di^5$ (b) $P \alpha n^3 /Di^5$

    (c) $P \alpha Di^5 /n^3$ (d) $P \alpha n^3 Di^5$

196. Sulfite formation rate is proportional to the rate of oxygen consumption   [   ]

    (a) $Kla = ½ dCso4/dt /C^*$ (b) $Kla = ½ C^* dcso4/dt$

    (c) $Kla = 2/C^* dCso4/dt$ (d) $Kla = 2 C^* dCso4/dt$

197. The dynamic method used in bioreactor with active cells is a   [   ]

    (a)   transition state method   (b)   steady state method

    (c)   unsteady state method   (d)   steady state method

198. The optimum product concentration may be selected based on   [   ]

    (a)   sulfur transfer   (b)   CO2 transfer

    (c)   O2 transfer   (d)   N2 transfer

199. If scale up is carried out by considering equal P/V then fermentation with increase in tip velocity, will be   [   ]

    (a)   more sensitive   (b)   less sensitive

    (c)   varies small   (d)   varies largely

Chapter 13

200. The sterilisable probes should withstand temperatures up to   [   ]

    (a) 50^0C   (b) 100^0C

    (c) 150^0C   (d) 121^0C

201. For measurement of intracellular metabolism frequently used method   [   ]

    (a)   IR   (b)   UV

    (c)   NMR   (d)   spectroscopic

202. Shaft power is given by strain gauge , value of potential de is given by   [   ]

    (a)   de = V/kst   (b)   de = Kst/V

    (c)   de = 1/VKst   (d)   de = VKst

203. The strain gauges for power measurement are mounted on shaft at   [   ]

    (a) 90^0 (b) 45^0

    (c) 60^0 (d) 30^0

204. The electrodes used for pH measurement have internal resistance   [   ]

    (a)   100-200 MΩ (b)   200-300 MΩ

    (c)   300-500 MΩ (d)   150-200 MΩ

205. In D.O measurement the response of membrane covered sensors is of order of   [   ]

    (a)   0-10 sec   (b)   0-5 sec

    (c)   5-10 sec   (d)   10-100 sec

206. Exit gas composition is analysed by   [   ]

    (a)   IR analyzer   (b)   UV analyzer

    (c)   γ ray analyzer   (d)   X-ray analyzer

207. Turbidity can be measured by   [   ]

    (a)   conductivity meter   (b)   strain gauge

    (c)   spectrophotometer   (d)   polarimeter

208. For maintaining sterility antifoam agents should be exposed to the temperatures of   [   ]

    (a) 100^0 C   (b) 90^0 C

    (c) 120^0 C   (d) 160^0 C

209. Proteins can be measured by   [   ]

    (a)   HPLC   (b)   GC

    (c)   polarimeter   (d)   ion exchange chromatography

Chapter 14

210. BOD is measured at   [   ]

    (a)   5 days, 20^0 C   (b)   3 days, 20^0 C

    (c)   7 days, 20^0 C   (d)   10 days, 20^0 C

211. In ETP primary treatment consists of   [   ]

    (a)   physical-biological treatment   (b)   chemical-biological treatment

    (c)   physical-chemical treatment   (d)   physical-aerobic treatment

212. In ETP secondary treatment consists of   [   ]

    (a)   physical treatment   (b)   biological treatment

    (c)   chemical treatment   (d)   chemical and biological treatment

213. Dissolved oxygen in aerobic biological reaction is kept around   [   ]

    (a)   2-3 mg/l   (b)   5-6 mg/l

    (c)   3-5 mg/l   (d)   5-10 mg/l

214. The optimum temperature for biodegradation is   [   ]

    (a) 20 ± 2^0C   (b) 15 ± 2^0C

    (c) 35 ± 2^0C   (d) 25 ± 2^0C

215. The biogas from anaerobic bioreactor contains   [   ]

    (a)   N2   (b)   CH4

    (c)   SO2   (d)   CO

Chapter 15

216. Tissue culture mainly deals with   [   ]

    (a)   skin cells   (b)   hair cells

    (c)   liver cells   (d)   heart cells

217. Gene therapy is closely related to   [   ]

    (a)   molecular engineering   (b)   metabolic engineering

    (c)   bioprocess engineering   (d)   tissue engineering

218. Synthetic polymer may be modified by adding   [   ]

    (a)   1-2 amino acids   (b)   1-3 amino acids

    (c)   2-3 amino acids   (d)   3-6 amino acids

219.   Retrovirus are encapsulated in   [   ]

    (a)   lipid layer   (b)   protein layer

    (c)   cytoplasmic layer   (d)   endoplasmic layer

220.   Stem cells are used in   [   ]

    (a)   for transplantation   (b)   for growth

    (c)   for injuries   (d)   industry

Key to MCQ

1 (d)	2 (b)	3 (b)	4 (d)	5 (a)	6 (c)	7 (d)	8 (c)	9 (b)	10 (c)
11 (a)	12 (b)	13 (c)	14 (b)	15 (a)	16 (d)	17 (b)	18 (c)	19 (d)	20 (b)
21 (c)	22 (b)	23 (b)	24 (d)	25 (c)	26 (a)	27 (b)	28 (c)	29 (c)	30 (b)
31 (c)	32 (a)	33 (b)	34 (c)	35 (c)	36 (a)	37 (b)	38 (a)	39 (b)	40 (a)
41 (b)	42 (a)	43 (b)	44 (b)	45 (b)	46 (b)	47 (b)	48 (a)	49 (b)	50 (c)
51 (b)	52 (c)	53 (a)	54 (b)	55 (b)	56 (a)	57 (b)	58 (c)	59 (b)	60 (c)
61 (a)	62 (b)	63 (b)	64 (b)	65 (b)	66 (c)	67 (c)	68 (b)	69 (b)	70 (c)
71 (b)	72 (c)	73 (c)	74 (b)	75 (c)	76 (a)	77 (b)	78 (c)	79 (a)	80 (c)
81 (b)	82 (b)	83 (c)	84 (b)	85 (c)	86 (b)	87 (a)	88 (b)	89 (a)	90 (b)
91 (c)	92 (b)	93 (b)	94 (a)	95 (b)	96 (a)	97 (c)	98 (d)	99 (c)	100 (c)
101 (c)	102 (c)	103 (b)	104 (a)	105 (b)	106 (b)	107 (c)	108 (b)	109 (b)	110 (a)
111 (b)	112 (c)	113 (b)	114 (a)	115 (a)	116 (b)	117 (a)	118 (b)	119 (a)	120 (b)
121 (b)	122 (a)	123 (b)	124 (c)	125 (b)	126 (a)	127 (c)	128 (a)	129 (c)	130 (b)
131 (b)	132 (a)	133 (b)	134 (b)	135 (a)	136 (c)	137 (b)	138 (b)	139 (d)	140 (b)
141 (a)	142 (b)	143 (b)	144 (a)	145 (b)	146 (d)	147 (a)	148 (c)	149 (b)	150 (a)
151 (a)	152 (c)	153 (b)	154 (c)	155 (c)	156 (b)	157 (a)	158 (b)	159 (a)	160 (d)
161 (b)	162 (c)	163 (d)	164 (a)	165 (b)	166 (c)	167 (c)	168 (b)	169 (c)	170 (d)
171 (b)	172 (a)	173 (b)	174 (a)	175 (c)	176 (d)	177 (c)	178 (b)	179 (b)	180 (c)
181 (b)	182 (c)	183 (c)	184 (d)	185 (a)	186 (a)	187 (d)	188 (b)	189 (b)	190 (b)
191 (a)	192 (c)	193 (b)	194 (c)	195 (d)	196 (a)	197 (d)	198 (c)	199 (b)	200 (d)
201 (c)	202 (d)	203 (b)	204 (c)	205 (d)	206 (a)	207 (c)	208 (d)	209 (a)	210 (a)

Index

D

Data lagging 237

Dead spaces 130

Debris 6, 171

Decay phase 39

Decimal reduction time 102, 105

Degree of dispersion 149

Del factor 110

Deoxyribonucleotide 14, 16

Descending curve 218

Deterministic value 106

Dextran 178, 179, 182

Diaphragm gage 220, 225

Diaphragm values 130

Dielectric constant 181

Differentiated cell 250

Diffusion transport 140

Digital panel meter 235

Dihydroxy acetone phosphate 27

Dilution rate 81, 85

Diphosphate 24, 27, 35

Discharge coefficient 163

Dispersion model 94

Dispersion number 94

Displacer 189

Dissociation extraction 176

Dissolved oxygen 121, 123, 136, 140, 217, 231

Distribution coefficient 173, 177

DNA molecule 16

Doubling time 8, 38

Down comer 77

Down stream 1, 136

Drag force 161, 163

Drying process 199

Dynamic model 79

Dynamic optimization techniques 234

E

Eadie–Hofsee plot 64

Effluent stream 83, 88

Electro dialysis 20

Electromagnetic waves 101

Electron acceptor 25

Electrophoresis 20

Enantiomer 49

Endogenous metabolism 38, 40, 42

Endoplasmic recticulum 10

Entrapment 46

Enzyme activity index 66

Enzyme catalysed reaction 59, 82, 98

Enzyme denaturation 72, 74

Enzymes 4, 27, 30, 32, 49, 56, 57, 71, 125

Enzymes inhibitors 66

Equilibrium coefficient 61

Ethylene di amine tetra acetic acid 66

Euglena 14

Eukaryotes 9, 14

Eulers constant 109

Exist age distribution 89, 91

Exit streams 217

Exponential growth 37, 38

Exponential growth phase 37

Exponential phase 36, 144

Extracellular matrix 246, 247

F

'F' curves 90

Facilitated diffusion 53

Fatty acids 27, 50